Small Shop Production of Custom Wood Doors

by
David R. Sochar

Big Oak Publishing

Copyright 2021 Big Oak Publishing

All rights Reserved. No part of this book may be reproduced, stored in a retrieval system, or transmitted in any form or by any means without the prior written permission of the publisher, except in the rare case of brief quotations embedded in critical articles or reviews.

ISBN 978-1-7367075-0-0

First Published September 2021

Published by
Big Oak Publishing
Ditch Road
Westfield, IN 46074

Edited by J.D. Ho
Cover designed by Courtney Mandryk

ACKNOWLEDGEMENTS

I wish to thank my wife, Cindy and my children, Kate and Ben, for their unwavering support in my efforts over the years. From holding the end of the board to full-time employment, to commiserating with me on the things that can get us all down, to celebrating the victories. They were there. Their love, their belief and smiling faces mean everything when the world seems to be conspiring. Thank you.

I wish to thank all the woodworkers who have shared their knowledge and added to my abilities over the years. I also wish to thank those I taught, as teaching, I found, makes one think clearly and with reason—qualities I can always use some help with. Their unselfish support has affected the work of a career and carried this book to its completion. Woodworkers are the best. Thanks, guys.

The Author

With over 50 years as a professional woodworker, the author has worked in or created several positions that fulfilled his need for challenging, interesting work. Building custom passage doors still holds his interest as it has for the bulk of his professional career; the variety, development, and execution of the passage doors keeps it all interesting. Never boring.

The author's interest in wood developed with model-building and slot-racing, then moved to include more wood than plastic or metal. When faced with a high school guidance counselor, it all came out: "I want to make wood furniture."

The counselor had no idea how to deal with that, so off to Big State U., where he languished for a few years, finding classes that gave a good background in basic design and some art history. In time, he found a few contacts who did a little carpentry to make an income. Well, the Business Administration degree was never going to happen anyway.

A few years repairing antiques and making picture frames helped narrow his interest and teach the basics. Antique restoration exposed him to all sorts of furniture forms, styles, and approaches to joinery and construction. A first mortise and tenon was made during this period, and a love affair began. The next move was to a small architectural shop that made movable and fixed slat louvers, mantels, doors, curved stairs, and stair parts.

All the shop hands and office staff were over 60 except for another young man, the lead stair-builder. No particle board. No plastic laminate. Experience was deep, and there was plenty to learn. He soaked it all up, asking enough questions to be a real pest, and questioning enough of the methods to realize there was more. Much more.

There is no simple or direct path to becoming an architectural woodworker. There are some schools now teaching peripheral classes, but almost all formal instruction today is furniture-making. Architectural woodwork is just not a single subject taught in the trade schools or academies. Strange indeed, since so many historical buildings, famous residences, and treasured architectural structures exhibit fine architectural woodwork in their execution. These same buildings offer great visual lessons for the woodworker who travels to visit and spend time luxuriating in the details that make the edifices so rewarding.

Moving from shop to shop for a couple of years, he realized he would never find the dream shop. He struggled to find a stable, supportive work home, but no employer could step up. Finally, in 1990, he just walked out of a corporate shop he had built up for the owner. While surveying a grim employment horizon, the phone rang, and then rang again. The spouse, the ever-supportive spouse, suggested raiding the savings account to buy new equipment if the phone was to be answered. After building a home shop in "the backyard" in Westfield, he was finally able to say yes when someone wanted something made.

Better materials and the best methods were central to the shop, and he soon was producing doors and other architectural items of high quality and good design. Often, he would try alternate methods or materials, making for a better product, but the additional time required meant the income was lagging. The solution was growth, also fueled by a desire for more equipment, better equipment, more people, and better benefits. The shop grew easily, requiring first part-time, then full-time help.

Eventually, Acorn Woodworks had to move from the country shop to lease space in 1996, and then to a larger purchased shop in 1999. The early years of shop set-up for others paid off with a well laid out shop, and with properly placed equipment that could build about anything. Curved staircases, large entry doors and frames, even 30-ft.-diameter round rooms—all could be built while other, more common items were also made.

Finally a place where process became central. Efficiency followed process. Productivity followed efficiency. Everyone's input was constantly sought. The best way? Best materials? Drop a step, add a step? How? Why?

The desire, the motivation to move forward and upward, was always there, always working, driving change from within. He questioned other professionals (his clients included highly successful surgeons, attorneys, and manufacturers): What is it that makes the status quo so undesirable? Do you find yourself exploring new avenues, new methods for your process also?

A new dialogue of a sort resulted. One that, even though it was non-verbal, would help develop Acorn products ever further, increasing the overall performance and real integrity to where it is said these products will easily last over 100 years.

This is the result of that journey, the lifetime of daily "cutting the board." A successful business, satisfied clientele, professional reputation, and engaged employees, are all a part of the shop. But, more than anything, the author created a venue in which his ideas of quality and process and excellence could live, develop, fail or succeed, and find a place in modern small shop custom door production.

Preface

For thousands of years, doors have served us. Protection, warmth, community, division...there are many reasons for doors, and many doors to satisfy the demands put upon them. Today, doors can be decorative as well as functional, secure as well as flimsy, handsome as well as plain. And much, much more. Doors today can not only keep the varmints out, but they can prevent bullets from piercing them. They can even repel the blast from nuclear weapons.

Why a book on making doors? There are hundreds—thousands—of books on woodworking. Surely the subject has been covered many times before. I felt that of the four to six currently available books on making doors, each book fell short somewhere. Variations of the basic door were missing, curves were missing, sidelights and transoms almost never got a mention, and tooling, technique, and real examples were non-existent. So this book had to be.

I write this book for other woodworkers. Not managers so much as bench-men and women. I can see it dog-eared and worn, dusty and a bit bent. Used. Out in the shop, being passed around. Not living quietly on a shelf. It is the book I wish I had come across when I was in my 20s. It would have saved me quite a bit of work. And my co-workers would have appreciated it, since I could then have stopped quizzing them so frequently. I think this book presents descriptions and solutions to the increasingly rare and arcane problems we face as designers,

architects, and others who continue to design things that fewer and fewer of us know how to make.

I heard a futurist speak in the 60s about how everyone in the next century will have heart disease. And since it will be so widespread, our ability to treat and even cure it will be highly developed and available everywhere at reasonable cost. No longer would it take a life's savings to save a life. The loss—the exchange—would be that some skills will be sacrificed. Skills like making a chair from a tree. What was once a semi-common trade will be replaced by heart surgeons, then heart techs. Perhaps like a Quik Lube Oil Change. And the chairmaker? He could become like a god. Or as prized as a heart surgeon of the 60s. His chairs would be evaluated, limited, critiqued, prized, traded, collected, and occasionally sat upon. While the plastic and metal chairs churned out by mass production gathered at our feet, used but not loved.

Where do I sign up? I was already moving towards learning to make furniture when I heard this. It certainly did strengthen my resolve. But a god? Nope. A heart surgeon? Nope.

But making doors has provided a very nice life for my contemporaries and I for over 50 years. The job satisfaction is off the chart. I would hope to share my experience with others in order that they may also find the pleasure I have found in this work. This niche field presents opportunity for many shops since doors, and especially entry systems, are a bit too large to ship easily, so they lend themselves to local manufacture.

Early on, I developed a deep respect for classic woodworking—18th-century furniture and design, *Fine Woodworking* magazine, which then showcased examples of craft and skill. Yet in the commercial world, I saw only careers that involved busting up plywood for boxes. I was fortunate to avoid that track, and I eventually made my own path.

This book is my attempt at preserving the skills and knowledge I have acquired, and passing it on to others so that the art and skill—and pleasure—of door-making is not lost, and so the world does not end up with only soulless plastic and metal doors.

As wood doors have fallen from the realm of woodwork manufacture, the process of making doors has fallen from the common instructional books. It is unlikely this occurred by design, just a niche that was easily overlooked. It is not covered in reprints of the old classroom books, much less in new texts.

Doors are unique in the woodworker's kit of products. Now that automobiles and small aircraft use almost no wood, doors are the most obvious common examples of moving woodwork that is left. Using basic frame and panel, a door can be as simple as four or five pieces of wood, joined at or near right angles, and made movable with hinges of some sort. Frames, multiple doors, transoms, sidelights, sashed work, and more grow from the simple frame and panel.

Since doors are a basic frame and panel at heart, it is imperative that we be very fluent in the creation of such a basic product. Every part counts. Every step matters. While the first frames may have challenges and shortcomings, with time one learns where to apply effort and where to relax, and the frame and panel begin to come together with much less stress.

Stock preparation is crucial, as "start flat, stay flat" is described and expanded upon. Best usage, cut lists, and layout are also covered. Basic stuff, but in light of making doors, even the basics must be reexamined to find out how each step plays a part in the whole. Joinery will be mostly mortise and tenon. This fundamental joint is not easy to make, so with time, it has morphed into a hundred—a thousand—easier to make joints. Cope and stick is often mentioned as a "mortise and tenon," but it is merely profiles fit to each other. Yes, it does increase glue surface—one of the functions of a mortise and tenon—but it does not give broad surfaces, and it cannot resist racking like a real tenon.

Dowels are often offered as an alternative to the venerable mortise and tenon. But they do not have the cross-sectional amount of wood that a decent tenon will have, so the joint cannot be as strong as mortise and tenon. Dowels also lack the ability to give the woodworker a bit of "forgiveness" if a joint is off by a smidge or two.

Biscuits and third members are loose "tenons" by a good stretch. But, again, not the real thing. They rely on the surrounding wood for their strength, rather than adding to it as our favorite joint will do. Their niche might be as alignment aids, which are always helpful. But they cannot do the job of a real mortise and tenon.

Custom passage door production implies that any and every door that comes along will be custom made to size, fit, and finish. Drawings must be made in scale, and perhaps full size. Numbers generated and crunched. Materials computed and ordered. Basics adhered to or left behind. Challenges discovered and resolved. All with the pursuit of profit at the center, for we all must live by our work.

Work satisfaction will also be covered throughout the book. Satisfaction from one's work is fundamental, but usually only as an ideal, since it is often the first goal that is set aside. Pain or hurriedness or poverty—or all and more—take its place and rob the work of its value. Rob it of the gift that good work, well performed and executed to the finish, offers as its own reward. Satisfaction set aside is satisfaction denied.

Profit and satisfaction can go hand in hand. A fair profit is planned for, just as a successful door project is planned for. Profit will help satisfaction, and satisfaction will help profit. This all makes for a healthy and happy work environment. When so much time is spent in a shop, it must be enjoyable time, profitable time, satisfying time. To accept less makes for drudgery in the shop.

If this book fulfills the goal, we might see small shops all over the country making a nice living, doing the good work of making doors. The future door-makers may find their career easier to define and exploit because of this book. And in generations, this book will record the process and inherent thoughts of an early 21st-century door-maker.

Table of Contents

The Author	v
Preface	vii

One — 3
Estimating	3
Drawing	6
Materials	9

Two — 13
Shop Space	13
Equipment	20
Process	28

Three — 33
Stock Selection	33
Face & Edge	37
Accurate S4S	42
Layout	43

Four — 47
Joinery	47
Tooling	54
Hand Tools	57
Glues	58

Five — 61
A Simple Door	61
A Difficult Material	62
A Simple Door II	64
A Coped Door	68
Sashed Doors	75
Bolection Doors	77

Six — 81
Panels	81
Glass	89
Muntins	95
Louvers	97

Seven — 107
Plank Doors	107
Flush Doors	111
Secondary Doors	115
Dutch Doors	118
Radius Plan Doors	120
Compound Curves	123
Large Doors	124

Eight — 127
Frames, Jambs, & Sills	127

Nine — 135
Curved Moldings	135
Curved Heads—Rails	139
Curved Muntins	143
Curved Frames	145

Ten — 149
Weatherstrip	149

Eleven — 159
Gates	159
Overhead Doors	163

Twelve — 165
Woven Wood Panels	165

Thirteen — 175
The Business	175
Innovations	180
Beadulator	181
Roper-ator	183
Warranties	184

Index	187

Chapter One: Estimating, Drawing, & Materials

Estimating

"Plan your work, work your plan," is an oldie but goodie. When young, I resisted planning my work since that meant work was in the plan, and I would rather not. Work or plan, that is. Once the realization that work was in the future, no matter what, plans became more important. Planning is the simplest of advice that will always pay off in the shop.

Generally, a door will come to you as a request for pricing. The general dimensions are given, but you will be responsible for coming up with everything you are not given. Thickness, stile and rail widths, panel thickness and type, materials, joinery, profiles, and price. As the expert on doors, it is assumed you can arrive at all the missing info. Ask questions if you like, but there is a certain freedom in not being given every parameter, allowing the maker some latitude in how the joints are made, in widths and dimensions of various members, and in overall arrangement of the finished product.

Estimating—generating a price based upon drawings and specifications—is an art in and of itself. For the type of work we do, it is crucial to get it right. Old estimators, now passed on, are still revered as legendary by their employers, co-workers, and competitors. The ability to look at a project and accurately (within 5%) gauge a price is a valuable skill. Time and reflection will help train the woodworker to be able to predict these costs in the future.

I once worked for a shop that would not price things ahead of time, only after they were built. I was shocked, as I could only think how little work this would provide, and even that work would produce angry customers, surprised at their bill. They did have a lot of angry customers, so we modified the process and learned to price ahead of time.

Devise a simple form that tracks the two things you sell—time and material. Every day, list the time spent directly on your project. Also list the materials you used to build the project. When you are done, tally the time and the cost of the materials. So, now you have two numbers—what you sold the project for, and the cost. As you get better at this process, you will find the two numbers inform each other and help you define what is successful and what is not. While one is tempted to think the difference in the two numbers is profit, there is much that comes into play before we have profit.

I cannot stress how important this step is. Tracking all your costs for a job is the most important thing you do on the business side. It is what enables you to be a woodworker because you can sell your work at a profit, put groceries on the table, expand your business, hire help, pay taxes—it all comes from knowing how long it takes to build a door. Track your time in hours, not dollars. And track your materials by board foot (B/F) or lineal foot (L/F), again, not dollars. Since the value of a dollar changes, it is not a reliable scale. And $3,300 worth of lumber does not tell you how much B/F there is without backing up with some math.

As for your labor—how much to charge? "Your labor" may be a mischaracterization. What we really need to know is the shop rate. This will be what is needed to pay yourself to do what you do. This number is driven by all your direct costs: payroll, lumber, glass, even shop incidentals like glue, filler, and sandpaper. Indirect costs also will go to build your shop rate: rent, electricity, phone, sharpening, insurance (Social Security and liability, plus unemployment when you add an employee), and more. It is important you catch all your costs, less important which pile they go into.

When you pay $1,000 for a pile of lumber, should you sell the lumber for $1,000? Nope. You need to mark up that lumber, since it is a place to make profit. It was your money that bought it, your knowledge of who to call, where to go, your place to store it—all help justify that markup. How much to mark it up? I have used 35% most of my working life, but previous shops often had arcane methods that were more alchemy than science. Pay $6.45 per B/F, then sell it for $8.71 per B/F. This lumber, now valued at $871.00 per 100 B/F is a direct cost when it goes into the project. There is no shame in marking up, unless you plan on "highway robbery." Marking up materials is an age-old, honorable activity, worthy and righteous. Doing it early, before the overall profit markup assures you there is some profit built into the job, even if the rest of it goes to hell.

Now we have direct materials, so we need to define direct labor. This is the amount of labor required to build a project. One hour or 1,000. Our estimate will tell us how much that amount is—or should be—in hours and dollars.

Now we must determine the indirect costs so we can build a shop rate. This will be electricity, phone, rent, insurance, shop supplies, and even payroll taxes. Some labor charges are overhead. Fixing the toilet is overhead. Making a door is not. Add these up for a month and divide by 160 hours (four weeks at 40 hours per week). This number is your indirect cost. Let's say it is $3,400 per month.

Your pay should be part of this cost. But if you hope to make a minimum of $25 per hour, then you will need to figure about $40 per hour to cover the taxes. This will be 40 x 160, or $6,400.00 a month.

If we add the $3,400 and the $6,400 per month, we have $9,800 per month. Divide by 160 hours, and we arrive at $61.25 per hour. This is our shop rate, as a minimum. Ostensibly, this will pay us, pay the bills, and keep us afloat until the next time.

What is missing is the fun—the profit. This is where we hope to grow, to make more money, to expand our product line, to reward employees. If we were to add 10% profit to our overhead number, we arrive at $67.38 per hour. If you can generate a solid 10% per year, after a few years, you will find bankers, con men, and "business brokers" wanting to loan you money. Better yet, you will be able to grow, add machinery and labor, reward yourself and others with bonuses, and adapt to the ever-changing world we work in.

Let's take a quick look at a sample project and see how it will work:

My estimate says it will take 100 hours to build. Direct labor will be $67.38 x 100 = $6,738.00 for the labor component. The estimate has the direct materials at $3,388.00 selling price (meaning we paid about $2,510.00). The total for the job will be $10,126.00. If there are to be any other taxes, like sales tax, they are added at the point of sale. Time and materials, pretty simple.

Write up your proposal at this point, and include as much information as you can. You want to present as complete a number as possible, even if it changes. If you want to present a price that is two-thirds what it will be, then, when the customer gets the bill, it does not jive with the proposal, and things go downhill from there.

If you determine you need more money from your efforts, that is fine (assuming the bank account can support it). Give yourself a raise (including taxes), and raise the overhead number by the same amount. As you get more accurate in your pricing, and it becomes more predictable, you learn where you can cut back if you need the work, and/or add on a bit if you know where you stand.

You may find getting the profit a bit elusive at first. No matter; that is normal. Stay the course, track your time, track your costs, refine the process. As you estimate jobs better, and produce known times in the shop, you will find yourself making money.

Once you have the work, it is time to revisit your materials budget to see what you will have to work with. Order materials judiciously—they can be very expensive, and the accurate estimate will start paying off immediately with lumber purchasing. Minimizing waste, and utilizing what you generated is one place a shop can earn solid savings.

After 40 years of estimating, I still have to grab a yellow pad to figure what any door might cost. I tend to focus on parts of the whole rather than the whole. I can easily say that for the last 12 years or so, the materials side of of our work is one-third the selling price, while the labor rate makes up two-thirds. This holds true in individual costs as well as annual reports.

About two years ago, our shop rate hit $100 after languishing at $85.00 or so. I had not calculated a 3% cost of living raise for years, and there just was not much cash in the bank, yet we were busy, working about 75 billable hours a week. That 3% makes a big difference, and we now find it easy to invest in new equipment or devices that will help our efforts.

The estimate is the core document that will make or break your business. Different folks may use different methods to figure overhead or generate pricing. What matters is that you develop a system that works for you, that is predictable, and that helps your business. If you are having a problem that is nor solvable by you, there is no shame, but do get help. Hire a CPA you can work with, or contact SCORE (score.org) for help where you need it. Business, at its core, is pretty simple. If you can work wood, you can work the business, too.

Drawing

Shop drawings are often required by your customer—the general contractor, the owner, the architect or other design professional, or even other trades. You will also require some sort of shop drawing for your own use in building the door/project. You will use this drawing and rely on it to build the project, and to get paid when the project is complete.

There is a third purpose for drawing: to problem solve. Those whom I have trained all rolled their eyes at my "Draw a cross section!" admonitions over the years. I encourage questions, and I equally encourage solutions. But a woodworker who can identify a problem and then solve it is solid gold. I have often said that we are problem solvers first, and woodworkers second. Our chosen field is non-standard enough to provide many opportunities to solve problems. And a good solution adds to the satisfaction that is a good part of our work.

In particular, cross sections will help you to understand how joints may come together—or not. A good cross section may well be worth thousands of words, yet be elegant and simple. How a hinge interacts with the throw of a door; how art glass will be a part of a door, yet be removable; how a curved muntin will intersect a straight muntin—all resolved by drawing a cross section or two.

A practical shop drawing will include an elevation of the door and related items like sidelights or transoms. A section should show details of the construction through panels and molding joints, glass panels and how they are set and sealed, internal construction, etc. Each page should be labeled with the project name, the date, the scale used, and other identifying marks. Modern software can produce a title bar similar to that found on architectural drawings.

CAD drawing is rapidly taking the place of hand drawn documents for use in the workshop and to inform the customer adequately. If a CAD system is used, become proficient enough to excel at your drawing as a part of communicating to your customer your commitment to excellence. CAD drawings often have a cold, or inhuman, look to them since they are not made by a human being. This is no excuse for leaving off info that may be critical. Be sure the drawings have all the info anyone would need to build the project.

CAD drawing programs can develop cut lists generated from the drawings. However, most CAD programs are designed to build cabinet boxes from man-made boards, using limited sizes and shapes. Trying to shoehorn a cabinet program into listing doors is not a good program. The act of making a cut list makes the project come one step closer to life for the maker. As parts are visualized and listed, the maker gets a good preview of what they are about to do.

I am just ahead of the curve for learning with computers. Several years ago, I determined to "go digital." I bought a leading CAD drawing program and took a few lessons. There was a lot to learn. I found out that: (a) I do not think in the same linear way a computer does, and (b) I had no desire to learn CAD drawing, and (c) it took me longer to draw a door than it did to build the same door. Now I know.

I am a strong proponent of hand drawn shop drawings. I use 1/8-in. grid graph paper, with the lightest aquamarine lines I can find, in a simple 8½x11 pad. Many word processing programs have templates for graph paper also. A good scale, several pencils of different weight leads, a compass, eraser shield, and erasers will do nicely.

The copier is used to make a few copies of the basic outline to use for trial drawings that are not intended to be finished or transmitted. Just so I can try a few layouts without drawing the entire thing from scratch. Once the drawing is complete, it is scanned into the computer and filed and/or distributed.

The resulting pages at 8½x11 can scan easily, and then be distributed in e-mails. The graph paper lines mostly go away when the documents are scanned. I can draw with a scale and pad of graph paper, produce an elevation, a cross section or two, then scan and distribute in just a couple of hours.

Drawing also is an invaluable tool for making doors since the cross sections explain the building of the door better than anything else. Accurate cross sections help you decide how things will happen, as well as how much material is needed, then how much time is expected to be used. This makes a case for at least some sort of drawing in the pricing stage to help you determine the correct time and material required to build the thing. Be sure you let it be known you are a professional

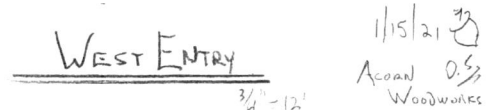

and expect to work in a milieu that is populated with other professionals. Some self-appointed "decorators" think that requesting a price and a drawing allows them to take that drawing and price shop it around town.

Distribute your pricing and drawing only to those who request it and have a need for it.

In fact, hand drawing also helps the visualization process as a preview of the cut listing. Drawing a project for shop drawings will start the process by forcing one to examine the sizes and functions of all the parts as they are put onto paper. Cut listing carries the visualization further, so that the actual cutting of the wood is merely the next logical step in this effort guided by process.

Drawings will need to be made for several reasons: to show the customer exactly what it is they are buying; to show yourself and the shop what it is that must be built; and to work out materials and construction details. Drawings are the means of communication you use to insure happy customers, profitable jobs, and continued success.

Be sure the shop drawings are accurate. In some types of construction, the door supplier will be asked to sign off on his work with a guarantee of some sort. The guarantee is backed up when the customer signs off on the door-maker's drawings. Many a loudmouth causing stress or consternation on a job has been silenced by the phrase: "These need your signature..." If you build what is signed off on, you are protected. There may still be a fight, but you can simply point to the drawings as your effective defense. That, and a project that is built like the drawings.

For some makers, it can be difficult to adhere to the shop drawings. Materials change, methods change, and while drawing, options may present themselves that may make for a better project in some aspect. These can be incorporated safely into the project by previously stating in the documentation that the "mill option" may be used for small changes that will not detract from the project. You may find it less intimidating to pick up the phone and make a call. Larger changes should be announced, preserving that safety sought from accurate documents. Deviation from the drawings is done at your own risk, and is not for vicarious fun. When one hangs out that figurative shingle, one should clearly understand the responsibility inherent with working for others. Security comes from the details you put into the drawings, so spend the time to make them nice and neat.

Another important component of planning your project is the specifications. Some projects—mostly commercial—will have a specification book that covers your work. Be sure you understand the specifications and any costs associated with them so it all is in your price. If there are no specifications given, then it is prudent for you to provide them. I learned years ago that it was in my favor to list the specifications when I quoted the work. No one else—large company or small—listed much of anything beyond species of wood.

Call out key dimensions; specify materials used and how they are to be used; and specify finish colors, part numbers, etc. This list of specs will help you in the shop in a few weeks or months when you need that info but do not have ready access to the job file, nor memory reliable enough to proceed.

Materials

First of all, buy wholesale. Introduce yourself to potential vendors and ask if you can do business with them. Be prepared to sign a personal guarantee of payment—standard operating procedure for small shop owners. Buying lumber is facilitated by knowing the "language" of the seller. Board foot (B/F: 1 in. by 12 in. by 12 in.), thousand board feet (MBF), moisture content, net tally (measure of lumber after drying), six-quarter, first and seconds, and many more terms will arise and affect your purchase. Know your terms, and don't be afraid to ask questions. Any decent lumber dealer can become a partner of sorts to you in your business. The relationship can—and should—be beneficial to both parties. As you both develop and grow your respective businesses, you will find them a valuable asset. Exploitive relationships, for either party, are neither stable nor productive, and are simply not good business. I have found most vendors to be of exceedingly good character—honest, fair, knowledgeable, and willing to help. Life is too short to have to put up with anything less.

Calculate the board footage for each piece and part of your project. Round up the fractions. Once you have a total, you need to add a waste factor. With hardwood grading rules developed to yield cuttings for the furniture industry, you can buy a whole truckload of 8/4 walnut, all on grade, and not get one piece clear enough to make a door stile. Think about the species you are going to use, and how it is likely to come to you. Some people double the calculated footage. Some cut it much closer, taking into account the species. Walnut will have a higher waste factor (I use about 80%) than poplar, where I might use as low as 35%. The relative size of the parts will also act upon the waste factor. A piece of 8/4 walnut, 14 in. wide x 3 ft. long may not exist in a truckload of firsts and seconds. But you will have no trouble getting 4-in.-wide parts. Is glue for width an option? How about thickness? Or even length?

A disturbing fact is that the very best hardwoods in my backyard will now never be available to me. The white oak, walnut, cherry, and others are all pillaged for the best boards to ship overseas. Longer lengths, wider cuttings, lack of defects, clean, clear grain, and figure are all factors that foreign companies will gladly pay for in advance. This wonderful local asset still provides jobs, income, and trade for many, but the very best is gone. Loss of those very best materials is crucial for the architectural woodworker since good length, width, and clarity are required for the projects they build. Without that material, prices are higher to cover the labor required to glue for width or thickness, or even change the project so it can be built.

When buying wood, buy from a vendor that is a professional wood dealer. This is how you insure you are buying wood that is legal, has been dried properly, and will do the job you are asking of it. Buy it in the rough after a discussion about the moisture content (MC). In the Midwest, we like the hardwoods to be from 7%–9% MC. A simple moisture meter will help you know what you are getting. It can also be used to measure that one really heavy board, the one you suspect never made its way through the kiln.

Your lumber dealer may be a distribution yard, collecting dried hardwoods and softwoods to offer their customers. Or it may be a mill. The mill will likely have a smaller selection of species, but better prices and materials for the species it saws. Be sure the mill is selling dried lumber. It is preferred to buy on net tally—the size of the lumber after drying. Green tally (before drying) can be as much as 10% higher since the wood has not been reduced in size by the loss of moisture. Selling on green tally is an indication you may not be dealing with the best lumber vendor.

Lumber can be bought green at a savings, and dried by the woodworker. Dry kilns can be simple—a solar kiln closely watched. Or the old reliable dehumidification kiln—basically a room with some vents, fans, a heat source, and a kiln schedule that will efficiently dry the wood without setting up stresses or other problems. Or a vacuum kiln that dries lumber from the inside out by drawing a vacuum in a sealed "bag," allowing the moisture to "boil" out. Even air drying is an option, but the lumber that is air dried will not reach the equilibrium moisture content (EMC) for stable wood use in interiors. Drying wood for one's own use is a good way to get your materials if you have a good resource nearby. Drying your own lumber also takes more time and effort than making a phone call, but can yield good results.

It pays to know the wood and your vendor. Know what causes internal checks, end checks, moving at the ripsaw, or moving after surface four sides (S4S). Know what is within your control and what is in the wood and unavoidable. Do not be afraid to question your vendor, but remember the best relationship with vendors is a partnership—both entities working together for better business all around.

The lumber should be in the "rough," rough sawn as from the sawmill. This will give you your best chance at getting the required thickness and getting the wood straight enough to use for doors. Store it in your shop if at all possible, so it is acclimated to an environment similar to where it will live.

Now it is time to make your cut list. This will be an extension of your estimate, which lists all the materials and labor it will take to build the project. The cut list will simply be a list of every part you need to cut out of the rough—or "bust out." Needless to say, an accurate cut list is integral to a good project. Plain and simple, though often pages long, your list of parts is as good a foundation as you can have.

List each piece by its size, with the largest—the stiles—first. The same format for entering the sizes should be used every time. First is quantity, then thickness, then width, then length. Use the rest of the line, if needed, for tenon length or other details. Develop a style or format for your cut list that is recognizable from cut list to cut list so the wheel does not need to be reinvented.

Working through the cut list, parts can be marked off as they are busted out. At times, it is prudent to list parts longer or wider than eventually needed. Acorn cut lists are almost always the final dimension, with additions for machining loss or end cuts. This accuracy is acquired as facility with the cut list and its functions, develops.

Commercial work will almost always require submittals and signatures so the various entities populating the job can all see and share the same info. Residential work is less formal, rarely requiring signatures, or even initials. In 30 years, I have required signatures on only three jobs, all of which were pretty fast and loose customers who did not seem to grasp what they were ordering. The signature was there to protect me in case they balked for some reason.

Years ago, I sent out a proposal as my first response. The proposal had a verbal description and some highlights: species, matching frame, hinges, weatherstrip.

Not too detailed—so as to avoid befuddling the customer. I realized after a while that I was the only one providing detailed proposals (things got more detailed over the years) along with pricing. The problem is, the customer—builder, contractor, or homeowner—wants to see a drawing first. They care little about a paragraph of specs. They want the price and a drawing.

Today, we do a detailed list of specifications that makes it all clear as day to see, for instance, what size and finish hinges we propose, as well as a scale drawing and an elevation. These are not line-item priced. No need for over-detail. The proposal is typed, legible, and packed up so it can easily be emailed and discussed over the phone, etc. This package is adequate for several things. It establishes the entire scope of work and who will do what part of it. It establishes the materials that will go into the work, as well as the design, look, and dimensions of the project. The package also presents you as a professional craftsman who has thought about all the details, and who anticipates a good project.

 **Acorn Woodworks**
David R. Sochar

Westfield, IN 46074
317-867-4377
davesochar@gmail.com

Customer Pat Redact   **Job** Euclid Ave

Specifications:

Species: Honduras Mahogany (sweitania macrophylla), pattern grade.

Items: Single Door at 1-3/4 x 32 x 86-1/2   4-5/8" wide exterior jamb

Secondary Species: None.

Hardware: 3 4-1/2" hinges, architectural grade, square corner, heavy duty. Without tips. Ball bearing

Latch: By others Mortise latch, add $247.00 Tubular latch, add $180.00

Handing: Right or Left

Sill Type: Oak sill, canted

Weatherstrip: Verticals and head horizontal – Force 5 Compressible foam, with corner blocks. Removable for finishing.

Glass: Code compliant, no logo, clear glass. Single glaze and insulated units are all set with either a light or dark RTV silicone sealant on both sides of the glass. Low E – Add $281.00 plus tax

Exterior Trim: Brick mold is a WM180 pattern, 2" wide. Typically shipped loose and long for field install.

Includes: Shop drawings, Delivery, machine and hand sanding, dummy hinges screws.

Excludes: Latch and latch prep (see above), install, or finish.

As above, Clear glass: $4,546.00 plus tax

Terms are 50% deposit to start work, balance due after delivery.

Fine Architectural Millwork
Bench Made Furniture

Chapter Two: Shop Space, Equipment, & Process

Shop Space

Shop space for doors needs to be a dedicated space. Professional or amateur or somewhere in between, makes no matter, a dedicated space is required for building doors. They take a lot of space, and if not provided for, doors will clog your shop and be a mess. Layout and handling must be roomy and easy, and not compromise your efforts at a smooth, well planned build. Too crowded, or too spread out—either makes for inefficient manufacture and loss of ability to turn a profit.

A solid bench is required, very flat and very stable, large enough to hold the door after it is assembled. The bench needs to be situated logically, usually in the middle of the shop, with clear and open access on all four sides. Room is needed for two people, a large door, clamps, clamp pads, glue, mallets, and maybe more.

My primary bench is made of 5/4 oak rippings from a rough day at the ripsaw. Glued for width, then for thickness, the bench weighs about 250 pounds, is 44 in. wide and 92 in. long, with drawers below, and a shelf below the drawers. An end vise was added recently, proving the old dog can learn something new. A second bench is made from corn crib lath sticks, again glued for width and thickness. It has 2x6 spruce-pine-fir (SPF) framing lumber as the legs, apron, and shelf. It, too, is rigid and heavy.

Any bench tasked with supporting a door (or any frame and panel) should be level and flat. When the door is assembled, glued, and clamped, the bench will provide a flat plane from which to reference the parts of the door and the completed door.

The flatness of the bench is tested by the two sticks–two lines method. Clamp a five-foot-long stick of equal dimensions to the top of each end of the bench. Center the five-foot lengths on the bench so the sticks overhang equally. Tie some lightweight thread to one end of one stick so that the thread is on top of the stick. Pull it lightly to the stick on the opposite end, opposite side. Wrap it around so it is taut, not deforming the sticks, and still on top of the stick. Take the free end, and go to the nearest end of a stick, and tie it off, again with the thread, again, on top of the stick. Then take the free end to the first stick (the only end with no thread). Wrap again, leaving the taut thread on top of the stick.

The thread will create a large X just above the center of the bench. Examine where the threads cross at the center of the X. They should just touch each other so that when the top string is gently lifted the other moves up by half the string's diameter. That indicates the four bench corners are co-planar with each other. If the string does not touch, then you will need to raise a leg of the bench with a shim to level it. If the string is tight against the other string, and pulling it down, a leg will still require shimming.

Shims should be stout, flat, and the same size as the foot of the bench. Once flat, it may pay to mark the floor so that if the bench is moved, it can easily be put back to "flat" again.

A "back bench" is highly recommended. A back bench runs parallel to the long axis of the bench, about four feet away, and has a top the same height as the bench, and is open above, with shelves below. It should have a central power strip with tools stored in the shelves, all plugged in and ready to use.

The worktop at the same height as the bench is to be left open as much as possible. This allows us to store parts, tools, clipboards—anything being used for the project is on the back bench instead of cluttering the primary bench. My favorite back benches were 24 in. deep, but they then can become a collector, and lose their efficacy. Eighteen inches is ideal—large enough to catch the hand tools and some parts, but not so large as to become a storage area.

The space under the bench is not to be wasted. Firstly, it should have a shelf or more for portable power tools and objects that are used frequently. A power strip should be centrally located so the power tools below can be plugged in for convenience, and the other tools or jigs set for easy access and storage. Drawers are immensely convenient—mine hold hand tools in two drawers, with the third dedicated to paper, drawings, pencils, batteries, etc.

The shop space as a whole can be thought of as the interior of one's head. Just as machines, hand tools, storage, and completed work all have a place in the shop, they all occupy space in our head. Both spaces need to be organized, clean, obvious, and healthy. Acorn uses the term "process" to define the series of steps any project must go through to arrive complete and correct at the end. Defining that process is often collaborative, but sometimes it is a solo effort.

Some process is obvious and well established. Some is unknown, so we have to devise that path: the steps, the checks, the choices we have at our disposal. Almost as important: When the project is complete, we will take a good look back to see if labor estimates were right, if material purchases were adequate, and if the process needs to be improved upon. The process plan is essential since our work is complicated—there are several steps involved in even the simplest tasks we do. Without a clear plan, we would succeed only at making sawdust.

Process is a partner on the path the successful project will take. Good process will be an aid to efficiency in the shop, while lack of process will not save even a hobbyist's birdhouse. After 50 years in as many as 12 shops, "process" has emerged as the one underlying intangible that makes any shop not only tolerable, but pleasant, and even uplifting. Indeed, this book-length effort is but an extension of the "process."

But the shop space also must have some real elements that bear upon our efforts. Good lighting, good air (both positive and negative), good space, and good storage are essential to the productive shop. Doors and their related components all follow somewhat the same path in the shop, physically moving from one area to another, with each step advancing further in the process.

Lighting must be effective, adequate, and efficient. Natural lighting should be maximized and made as available as possible. Some areas, like over a bench or shaper, will require a higher amount of candlepower than the storage areas. Smaller shops can use the newer T-8 bulbs in 4-ft fixtures and place them easily as needed.

As I write, LED "fluorescent" tubes are replacing the "T" bulbs. I suggest using more than you think you will need. If for no other reason than that aging, which we all do, will affect our ability to see details. A call to your electric utility should supply info on energy efficient fixtures and adequate lighting. They might even have a program where they buy back outmoded fixtures or hand out newer ones.

Shops with plenty of height may elect to use larger high-bay or low-bay type fixtures. One must do the calculations to ensure that enough light is made available, as this type of lighting is not easily moved about. The high-

or low-bay lighting is efficient, but it is easy to go astray here. Such lights can be had in various color temperatures, from warm to cold, so be sure you get the color you want, in the range you need for seeing the wood in its "true" colors.

This is a dusty business, so good ventilation must be considered. A fan or two on stands or on a wall, can help keep air from getting stagnant. Windows and doors that open are a good way to get fresh air into the shop. Fresh air has no effect on the wood, but is a good way to help keep the woodworker happy and effective.

In fact, good health depends upon good air in the shop. Dust collection is important, and even life-saving, over time. Research continues to show that even small particles can cause disease, with length of exposure also a major factor. As a result, HEPA filters are now being marketed for small shops. Any good air defense will include more than one strategy.

Dust collection starts at the machines that make dust. Festool and Mirka changed the way U.S. woodworkers sand wood with their integral vacuum systems, which collect about 95% of the dust they generate. The vacuums can be connected to circular saws, routers, and other portable power tools to collect the dust at the source. The small diameter hoses are not suitable for use with stationary machines, so other avenues are to be employed.

Central dust collection is mandatory in a small shop looking to work productively and efficiently as well as healthily. Collecting the shavings and dust as they are made limits exposure and keeps things clean, saving sweeping time and the exposure associated with cleanup. Fixed rigid ductwork with various diameters, and gated to regulate flow, can carry all the shavings to a central collection point. The better collection systems will use a filtration system of bags or pleated filters before circulating the now cleaned air back into the shop. Therefore, heated or air conditioned air sucked into the system is returned at about the same temperature, relieving unnecessary strain on the shop HVAC. Acorn has a separate room for the fan, motor, and cyclone separator that allows the chips and most of the dust to fall out into a 100 cubic foot room that is cleaned out from the outside of the shop. This system, designed by Oneida, has been in place since 1990, with no flaws whatever.

Prior to the current system, a good portable unit was used, with 2-in. to 4-in. hoses and a plastic bag for the shavings. The planer could fill the bag in 15 minutes or less, so when doing a lot of planing, the bag management took as much time as the planing. Then, if I needed to joint a few edges, I had to move the hose. Ditto for the table saw, then the

shaper. And when changing and emptying the bags, one is exposed to a very high dose of fine particulate matter—exactly what one wishes to avoid.

Moving the hoses and handling the bags were all detriments that had nothing directly to do with building good woodwork—process. Process dictates that events in the shop can happen on their own when possible.

Good process means more than one thing is going on. Dust is collected and filtered, and clean air is returned to the shop, all while lowering the workers' exposure to unhealthy air and making for a cleaner shop. This makes for very good process. Good dust collection is almost invisible when properly fit into a shop.

The better companies selling duct collection systems will gladly plan a good working system for you if you supply them with a floor plan of your equipment with the required cubic feet per minute (CFM) for each machine. They will respond with a drawing and a list of parts and prices for the system they recommend. I strongly urge using a professional vendor for the design of the system. Good dust collection is expensive, but no more so than bad dust collection. Especially when the health of the occupants of the shop is considered.

Another form of shop air is compressed air. In its simplest form, it is a pancake compressor in the corner used for the pinner or a blow gun. The efficient shop will require air lines throughout the building—at each machine station and at several other centralized points. Indeed, some equipment requires air pressure to operate, so the compressor and lines must take that into account also. The air lines are similar to the dust collection lines in that they have limitations on the amount of air that can be moved.

The better compressors are two-stage, with rotary compressors being a step up from two-stage. The small shop would not normally use more than a two-stage compressor, unless they did a lot of sanding with air-operated sanders. The rotary compressors will outpace several sanders and still supply the shop with good air where needed.

Air lines can be safely made from a variety of pipes; a bit of research can reveal suitable lines as well as fixtures. Some shops, such as Acorn, merely use 1/2-in. flexible hose for the lines as well as for air supply to the equipment. This is easy to change if needed, and also easy to replace if damaged. We don't have runs longer than 30 feet. In a small shop that does not call for large loads of compressed air, small lines can be run efficiently. The air must be regulated to avoid damaging tools with air limitations on them. The air should have a water trap and drain, as well as a compressor drain. More advanced systems have auto-drains that open when the unit is turned off for the day. Such systems might also require a dryer for air-operated equipment like sanders or edge-banders. Water can build up in the lines, damaging tools and valves, and even spraying water on finely sanded wood or bare cast iron.

Good compressed air becomes a part of the process in a shop since it is relied upon every day to operate when called upon. If process is a tool, a method, then compressed air is support to that tool, essential for the proper, best operation of that tool, of that shop.

Shop furniture includes sawhorses and carts. An effort should be made to make all your benches, tall carts, sawhorses, and major equipment the exact same height. This allows for moving large and heavy items—like doors—from place to place without lifting. We use our four sawhorses every day, and the cart that is our "roll-around" has a 24-in. by 28-in. top that can carry up to 300 pounds. The roll-around has about five pancake drawers that hold wrenches, setting gauges, specialized tools, and electrical things.

Lastly, a productive shop needs a loading dock or specified area for unloading lumber, and loading out completed work. Most shops default to have both incoming and outgoing in the same area. If it is not crowded, and if it is easy to move about, it may work fine. But if some rough lumber—900 pounds on a cart—contacts a fine finished bit of work, it is the finished work that will suffer.

It is not unusual to handle large sidelight frames with sash installed that weigh over 200 pounds. Or pivot doors weighing 280 pounds. An overhead conveyance—a trolley, track, and winch—is an ideal way to lift the projects and get them out of the door. A used fork truck can be had for a fair price and will enamor management with the shop employees. That is a win–win all around. Employees will be grateful, and avoiding worker's compensation claims is a worthy goal.

Equipment

The equipment in my shop will be different from that in your shop. A third shop will also be different. Just as the products we make will differ from each other in their details. But we will all have a way to cut the rails to length, then to tenon them. The stiles will need to be mortised by any method you deem as good as you have. The important thing is that your equipment is in good shape, with sharp tooling and trustworthy settings. Most woodworkers have a list of equipment they would buy next given the chance. Mortisers, tenoners, shapers, and sanders all play a part in making a good door, but none is as important as the maker.

A well-set-up shop and bench will become a tool in itself—a door-making tool of a sort. The process becomes visible, and putting your materials into the process becomes second nature. Everything gets treated the same way—bust it out oversize, face and edge, plane to S4S, crown, layout, make joints, haunch and dry fit, then glue and assemble. A process that is always the same, but also different. A process that, if planned and respected, will result in predictable results every time.

It is not imperative that today's door shop have all the bells and whistles available to the woodworker. Shops can be primitive and shops can be luxurious. No matter—the process is similar. The equipment should be kept in top operating condition so it can perform at its outer limits without breaking down. Since doors are so large—and getting larger—they are often on the edge of the capacity of the available machinery. Hobby woodwork equipment is not sufficient for regular door production since it is too lightweight and just not designed to handle thick, heavy stock all day, every day.

It may be wise to think of the equipment as an impermanent solution. You will acquire a larger shaper, perhaps, or a wider joiner, or a better mortiser. So when you set out to build doors, you can do so with the best equipment you have today, knowing that tomorrow will bring improvements. The equipment will evolve according to need and availability.

Acorn long ago adopted a strategy that has paid off well over the years. All the benches and the sawhorses are the same height as the table saw: 36½ in. to be exact. This means that heavy or cumbersome parts can be slid about the shop with minimal effort. Nearly complete projects or ones with fragile faces, can be slid onto a mover's quilt and then moved around easily. We have four "T-horses" that are about 40 in. wide, and shaped like an inverted letter T, with the base on the floor and two stiles supporting the two top rails at 36½ in. These horses work all day, every day, supporting the end of a door, holding lumber as it is busted out, acting as an extension of the bench for swinging longer radii, and more. The next shop (?) will have everything at the same height, facilitating movement and nearly eliminating dead lifts.

Whether it is one door or a hundred, repeatable accuracy is essential to good door production.

Used equipment is available through sales and auctions, and can be had for pennies on the dollar when compared to new. In fact, older American iron machinery is considered by many to be superior to the foreign-produced offerings of today. The older, proven designs and the fact that most machines do a very simple job makes for good buys in used industrial equipment. Even three-phase power is available to the modern shop with rotary phase generators.

Hand held power tools also play a part in the production of doors. If for nothing more than the old argument, "wood to the tool, or tool to the wood?" Often doors and their components are so large that the tool is better taken to the door than the door fed through the machine. In one shop, two men used a large panel saw for sizing doors to length. In another, a track-guided circular saw was taken to the work for sizing doors. One method might be considered better or favored, but both work, and work well, as long as the limitations are considered.

Occasionally, the work calls for, or even demands, an investment in larger or more capable equipment. Usually this comes with repeatability, and the added machine will add accuracy, efficiency, or even safety to the process. Speeding up process is always a goal, but there are many other goals along the way.

Not the least is safety. Safety is implied as a goal, but needs to be addressed. In fact, it needs to be a constant, so the worker can assess each and every operation in light of safety. One of the drawbacks to pushing large boards through small machines is that the machine will eventually be overwhelmed by the process and will self-destruct mechanically. Or electrical limitations may surface and the motor or controls fail. This is the time to shop for a replacement. The prepared woodworker will have a list, written or imaginary, of the "next" few purchases. In fact, it is not unusual to find a woodworker browsing the auctions and sale listings online. Knowing what is available and approximately what it costs is valuable information to the shop that finds itself in need of key equipment.

If you need a machine upgrade but are unsure of what the machine will do for your process, give some thought to how much time the machine will save in a day, week, or month. Come up with one factor and extrapolate. Those are shop hours, priced at your normal shop rate. The goal is to have the machine pay for itself within five years. If your shop rate is $90 per hour, and by your best guess, it will save you about four hours per month, that is $360 per month savings. In five years, that is a whopping $21,600. If the machine costs less than that, then you should have no trouble paying for the machine. Or is it that the machine pays for itself?

Buying used equipment is often done without the satisfaction of kicking tires, since it is not close enough to examine. It is wise to budget enough extra cash to rebuild the machine or replace bearings, or perhaps a motor. When buying (and selling) machinery online in the past, Acorn had good experiences buying sight unseen, but some further expense was expected and seen with replacement transformers, or new bearings throughout. Going through a "new to you" used machine is a fine way to get to know the machine and its capabilities.

Acorn's equipment, January 2020:

- joiner: 8-in. generic brand, 2 hp
- planer: 18x6 Powermatic 180, single phase 5 hp
- tenoner: Powermatic 2A, 1961 model, with Accurate DROs on the two tenon heads
- mortiser, large: Maka DB-6 with a rebuilt head and Accurate DRO on the head, 1978 model, made in "Western Germany"
- mortiser, small—Powermatic 719-A, Makita chain saw mortiser, portable
- shaper, large—SCMI T-130, 9 hp, 3-phase with Accurate DRO
- shaper, small—Rockwell 43-340, 1950 model
- router table—shop built, 200 lbs., 30 in. by 40 in. top with Benchcraft router mechanism and Porter-Cable 3 hp router
- table saw—Delta Unisaw, 1992 model with Accurate DRO on the fence
- band saw—Delta 14-in. with 3 hp motor (!)
- vacuum bag and pump—single stage, 6 CFM pump with several bags and platens, typically knocked down
- sanding—two Festool 6-in. dual-action sanders and vacuums; assorted smaller sanders for specialty applications
- dust collection—Oneida fan, blower, cyclone, and ductwork, circa 1992
- hand planes, spokeshaves, chisels, carving tools, electric and hand drills, and much more

The two key machines missing are a large joiner—12 in. by 7 ft. or more is ideal—and a wide-belt feed-through sander. We have worked with the smaller joiner—two sets of bearings to be exact—for 30 years, and have built hundreds of doors with perfectly straight or crowned stiles. It takes a bit more time, and a good eye and hand, but it works well and takes up little space. Yes, an 18-in. Martin long-bed joiner would be ideal, but then things might just be too easy.

Acorn once had a wide-belt in a larger building. A 37-in. single combination-head Bütfering worked well from the day it was set up. Tuned up, one could control how much of a pencil line would be sanded off in a single pass. We built doors about 0.040 in. thicker than the finish size, and took off 0.020 in. on each side in two to four passes. This worked well. It leveled all the joints perfectly and gave a good finish to the doors. However, as with any wide-belt, the rails that went through cross-grained had heavy cross-sanding that took about 30 minutes or more per side to remove.

This turned out to be the fatal flaw on doors through the sander. The cross-sanding made the door harder to sand properly and took more time than if we had just sanded with the Festool sanders—no wide-belt. We still used the machine for many things, and it proved a reliable ally in the ongoing effort to control the wood.

The joiner is a simple machine consisting of two tables co-planar to each other, with the in-feed table adjustable for depth of cut or the difference between the surface of one table and the other. A fence holds the work at 90° to the table, or at an angle we need to machine on the edge adjacent to the face.

First: semantics. The machine can be called a joiner, or a jointer. A person who makes joints and uses a joiner is a joiner—noun. A machine is called either a joiner or a jointer—both nouns. To put two pieces of wood together is to join them—a verb. And the act of making the joint and then assembling it is jointing the parts together—verb.

Longer and wider are two qualities it is nice to maximize when considering a joiner, since it facilitates longer and wider lumber. However, machine expense and shop space are considerations. We use an 8 in. by 6 ft. generic import that functions very well. I dream of a 12 in. (or wider) by 8 ft. to 10 ft.-long machine that would joint lumber all day in just one pass.

The small machine does a fine job as long as you "dance" with it a bit. Hit the high part, then again. Then swap ends and try the length. Give it an eye after that pass, and hit the high part again. Check thickness. Repeat. Two to five passes for each longer board in place of one or two passes for a larger machine. Tedious, but we get good at it, and can go as fast as reason allows.

Using the rule of thumb, a new machine would easily pay for itself in less than the five-year standard of machine payback. That is, if the labor-dollar savings will save enough to buy the machine in five years or less, then it is worthwhile to buy, and it will pay its way. I have had some machines pay for themselves in less than a year with increased productivity (four-head planer, wide-belt sander).

A planer does a lot of hard, mundane work. Five hp might be considered a minimum, although I built doors with a 3-hp "four poster" that was only 15 in. wide. The Powermatic 180 we use today is a solid little workhorse. Eighteen inches in width by 6 in. maximum opening, it will plane down to 1/8 in. with good results. Variable feed speed helps. This machine once had a grinder attachment that bolted onto the top of the machine to sharpen the knives in the head. Very convenient, but the grinder made for a generous investment in time, made a mess of steel fragments everywhere, and did no better a job than the service we normally use to sharpen our tooling.

Every planer develops characteristics that are recognized once we work with them long enough. Usually, these are feed problems, and they can be remedied by going through the manual and setting up as best as you can, using the tips you can get from calls in to support. This Powermatic has a few quirks, but can easily go from hogging 3/8 in. off of cedar beams two or three at a time, to 1/8 in. finish planed.

The Powermatic we have is close to the last planer we will ever need. Wider would be nice. Thicker would be good. Beyond that, there is more.... In a larger shop, I had a Quattromat—a four-head planer/molder that could joint a face and an edge, and then plane all four sides simultaneously. Yes, perfectly straightened, squared, and dimensioned stock in one pass, at 30 feet per minute. The dimensions were set by digital counters, so it took seconds to set the size, and then feed parts. It was fast enough and accurate enough that it could easily be used for just one or two boards.

Equipment evolves. The range is visible for each and every piece of machinery one may need.

The shaper is another key machine in the shop. Real shapers have 1¼-in. spindles—or larger—and variable speeds, usually from 4,500 to 10,000 rpm. A generous-sized table with a solid and heavy fence is important. The fences must be solid enough to resist warping or deflecting as materials pass by. Speeds should be easy to change.

Some shapers have an electric switch to reverse the spindle rotation. When reversing the rotation, the spindle nut must be locked in place to prevent the tooling rotation from unwinding the nut. Use the groove milled into the side of the spindle in concert with a special washer that has a tongue that fits into the spindle groove, preventing the nut from loosening while the machine is running.

The work can then be fed in from the opposite direction and "climb-cut," or cut with the rotation of the spindle. This is an advanced and potentially dangerous technique that needs to be taught in person. The larger parts and smaller cuts are better for climbing than the heavy cuts and small parts. The reason to climb-cut is to clean up a cut. With some woods and some cutters, there can be tear-out along a rabbeted edge. The more brittle the wood, the greater the tear-out. Climb-cutting cuts the wood "before it knows it is being cut," as one colleague put it.

The large shaper we use can run split collar knives, matched tooling sets, 4-in.-diameter molder heads, and larger tenon discs up to 10 in., as well as rabbeting heads and adjustable angle cutters.

A power feeder makes the shaper much safer and adds versatility. By its nature, with three or four wheels on the work, holding the wood blanks securely on three sides, the feeder can make a simple molding job go quickly and smoothly. In fact, the feeder can be useful on almost any piece of equipment: table saws, joiners, router tables, and more. Typical feeders have two speeds, with two more in each direction. Feeders are available with belt drives, but the belts glaze over and do not hold the work well. Better yet, a feeder type is available with 6, 10, or 14 smaller diameter wheels that are especially good at holding the work.

The apparatus that holds the feeder in place is the remarkable part of the tool. The feeder can be rotated, turned 90°, placed any way you can imagine, and tightened down to feed your parts. Each point can be extended or retracted, and set so all the wheels are on the work, holding it firmly to the machine table, pulling them tight to the fence and table for accurate registration.

Feeders are helpful in establishing the "endless board." The endless board is a production term I coined for parts going through a machine as efficiently as possible. Butted end to end, all stock feeds better.

A feeder on a shaper turns a semi-threatening machine into the safest in the shop. The ominous cutter head with protruding knives will be covered on three sides by the machine fence and table. The fourth side will be covered by the feeder, preventing hands from getting involved.

Banging out a mortise by hand is an important exercise for every woodworker. A stout mortise chisel, mallet, and a block of hardwood will teach quickly how nice it is to have power equipment. The size and type of mortises we need for doors rule out handwork except in details or remedy work.

Mortisers (powered) come in several types and many sizes. The hollow chisel mortiser is the most common today, with a rotating bit inside a hollow chisel. The chisel and bit are driven into the wood by a lever or foot pedal on larger units. The apparatus looks like the cousin to a drill press, and indeed, there are conversion kits that can turn a drill press into a mortiser.

The mortise it makes is a bit rough on the inside mortise walls, but not objectionably so. Several mortises are plunged in a row for wider tenons. Benchtop hollow chisel mortisers are suitable for lighter and smaller work than we will be doing—except for one task: They excel at placing the mortise for muntins in the stiles and rails.

These are typically small mortises of 1/2 in. by 3/4 in. for the fixing of tenons in place.

Larger hollow chisel mortisers are available from large benchtop size to floor models, old iron. The older, larger hollow chisel mortisers had a foot pedal to drive the chisel in, while some featured powered drive and retract for the chisels. These can still be had on the used market, and easily put back into serviceable condition.

Another older style mortiser is the chainsaw mortiser. Just as the name suggests, a chain rides on a bar that is held vertically and driven into the wood. The chains come in different sizes for different mortise widths. These mortisers are fast and aggressive and capable of accurate work.

A variant of the chainsaw mortiser uses the same principles, but is portable. European and Japanese makers have developed portable chainsaw mortises for timber work. These are handy and surprisingly accurate while remaining versatile. A good example of taking the tool to the wood.

A larger and increasingly rare machine is the oscillating chisel mortiser, as made by Maka until about 20 years ago. These are heavily built machines that drive a cutter in an oscillating pattern that cuts the wood on one side of the cutting path ellipse, then retracts, ejecting the chips on the other side of the oscillation. This all happens very fast, and the machine can easily place a mortise 3/4 in. by 2 in. x 3 in. deep in hardwood in 10 seconds or less. The side walls are very smooth, and all four sides are square to each other, with a slightly scalloped bottom. One should not count on the bottom of a mortise for structural holding, so the scallops are immaterial.

While oscillating chisel mortisers are hard to find, they can be rebuilt and mostly need pneumatic attention since the machine controls are air driven. They can work all day and never break a sweat, despite the difficult cuts they make.

The oscillating chisel mortisers are being replaced by machine manufacturing with computer numerical control (CNC) routers. These machines are capable of remarkable accuracy and repeatability. Software and cutter technology is evolving to make these machines more useful at making mortises.

A simple and very effective type of mortiser is a 3-hp plunge router with guide rails fixed on the base. The stiles can be gang-clamped and spaced onto the bench, and mortises for several doors can be made in one run. Use a solid carbide up-cut spiral bit, as long as you can find, and set up to make several passes. While most bits have only a 2-in. cutting length on a 4-in. or longer bit, repeated plunges can make mortises that are 3½ in. deep.

One floor model of a mortiser is actually a plunge router in a fixture that is used for clamping and holding the wood parts. Another variant uses a proprietary motor set up to spin the bit, with clamping and holding methods for the machine.

And yet another type of mortiser is the smaller oscillating bit type. Many of these feature two tables for clamping parts so they can be run two at a time. These have rotating bits of different diameters for the mortise, and depth and length are controlled by the operator as it is set up. These produce small, round-ended mortises suitable for furniture shops and in production settings.

A variant of that type of mortiser is the Festool mortising machine that links the mortise it makes to a loose tenon. These are portable tools, but they lack the versatility needed for door work.

Tenons can also be made several ways with universal machines or more dedicated machinery. The simplest way to make a tenon is with a table saw. Saw in the shoulders, then the cheeks. Four cuts, and you have a reasonably accurate and serviceable tenon. Standing the rails on end for a cheek cut is much easier, safer, and more accurate with a tenon jig of some sort that will hold the part vertically while the jig is fed into the saw to make the cheek cuts.

Another way to make the tenons is with a dedicated tenoner. For many years, Powermatic made the 2A with two shafts making the cheeks of the tenon and shoulders. A cut-off saw and some machines had the optional two vertical shafts to make top and bottom copes. These machines are relatively simple and can be rebuilt easily and put into service. Other old iron like Yates, Wadkin, and Oliver all made two-to-three-station tenoners. A few offshore makers have copied the old iron and are shipping machines into the U.S. based on that older style.

European tenoners have evolved over the years and now have several stations/spindles, which are each capable of spinning larger diameters of tooling, with several sets of tooling on each spindle. The spindle height is numerically located for accuracy and virtually no setup time. Each spindle can be loaded with four to six different cope patterns or panel raises.

Heavy shapers can also make tenons with disc-type tooling. The cutting geometry of shaper-made tenons produces quite a bit of cutting resistance, so be sure to seek out tenon discs designed to cut tenons on shapers. Some of the larger shapers have tenon tables with half the machine table unlocking and sliding past the spindle. Others have bolt-on fixtures for holding parts. You can make one or have it made. The commercially available tenon jigs are not heavy enough for passage doors. Any jig you make or have made needs to be heavy enough to not deflect or have any give.

A 3-in. long tenon on a shaper will require tenon discs almost 9 in. in diameter. Be sure your machine can safely spin tooling of that size. Shapers come with no instructions for a reason—you are expected to be able to make jigs and develop guards that keep you safe on a shaper. Tenoning with large discs calls for a heightened sense of safety so you can predict and resolve problems before they *become* problems. Even if you don't prepare yourself, once those heads start whirling around at 6,000 rpm, they will command your attention.

The shaper can be used to make copes on a tenon made on a square shoulder tenoner or table saw. This is a good way to add coped joinery as your tooling develops. We use the Powermatic 2A to cut tenons accurately, just a bit oversize. Then cut the copes on the shaper with tooling that cuts all surfaces and trues up everything to the tolerances we like.

It is not unusual to adjust a tenon's fit by 0.020 in. or less. This can change a mallet fit to a sliding fit. Another 0.020 in. would make it a sloppy fit. Consider the number of joints, as well as the glue, when you are setting up (before you accept a fit). A few joints, each a little tight, can be clamped together, and a fast-setting yellow glue can be used. If there are many parts, all tenoned and mallet-tight, then there will be a sweat-inducing mallet-flailing maelstrom trying to get everything pulled up. Better to have loosened it all up a bit and enjoyed the assembly instead of sweating it out.

The equipment in a shop will help dictate what can be made in the shop easily and handily. If you work at the equipment's limitations—or beyond—then the products are destined to be less than they could be, too time intensive to sell, and the marginal work will prevent the shop from enjoying any measure of success. There is a bit of chicken/egg in this: Does the product determine the equipment needed, or does the equipment determine the product? The reality is the evolving environment in which we work, so as we add new equipment, it will be more suited to the door work. Same with tooling and other purchases. Allow the door production to determine your path, your decision-making. This path, this set of operations, same and different, becomes the process.

As time moves along, you will seek out replacement equipment if, for no other reason than that the existing is worn out. Looking back, the replacements will in all likelihood show the evolution in the shop's abilities and productivity.

Every shop will equip itself differently, and solve the problems we all encounter their own way. Some may be more productive than others, some may reflect personal preferences, some may be sloppy. Vive la différence. But every shop committed to making good quality work will evolve over the years.

Process

Process is an abstract value I feel is important to the success of any shop. Process is the act of moving the job through various standpoints or operations to a goal. Particularly with door production, where nearly every type of work is the same—or similar—yet everything is different in every product. Every frame and panel door will get mortise and tenon joinery, but it may be unique when compared to all the other mortise and tenon joints before or after.

Despite the fact that mortise and tenon joints will differ in some aspects, nearly every day, they are—always—mortise and tenon joints. Same but different. Process will be the same, and for the sake of efficiency, will be

the same every time. No need to reinvent here. The challenge is to determine what the joint will be, and then to simply make it. First the math: Tenons should be about one-third the thickness of the stock. They typically run about 1¾ in. deep (before cope), and they will run the width of the rail, less a haunch, if anything. The process can be thought of as a framework that exists to help fill out the details, to give form to the abstract.

The process is everywhere as a result. Stressing at the joiner, wondering if that kinky white oak will face out, process dictates the best effort. Facing it will be the only way to know if it will suffice. Since it is a lamination in a thicker assembly, does it need to be faced 100%, or will a nominal facing work as a good reference with which to plane? Process is running the show, firmly suggesting the path to take. Every day, every step, all the shop work is guided by the process. Guided to the end, as the project is loaded and delivered.

Process also takes on a larger role when it helps guide major decisions. Some years ago, Acorn was faced with the nasty realization that the glue we were using—Titebond III—was failing in west-facing doors. These doors all had very dark or black finishes and received the rising sun flat to the face. The panels that were glued for width had open joints on the exterior panel face, running down into the thickness of the panel. Most of the joints in the eight to ten panels we replaced were still intact on the interior side, giving no hint of releasing.

We examined process. Sharp joiner knives? Clamp pressure? Fresh glue? Fresh joints? All were positive, all operational, the process was intact. Nothing out of the ordinary. A phone call to the Titebond tech department cleared up the dilemma. The glue simply loses its strength when heated. The surface temperature of the doors that had failures could climb to over 190° in clear, full sun. This made the glue lose enough grip that it allowed the joints to open.

In all of the failed panels, the glue joint failed only on the exterior side, perhaps two-thirds of the way through the thickness of the panel. If the panel did not heat up on the interior side (no sun), then any joint stayed tight on the inside.

While process was intact, and followed, it clearly needed to include this new information. The first thing to do was to find a workable solution. We could not tell the customers they had finished the doors incorrectly—we had not warned them against dark finishes, and we preferred to not have to police that for the future. Change glue? To what? We selected Titebond III because it was exterior rated, easy to apply, and affordable. Stop gluing for width? This was not practical, though it eventually led to the solution.

First, we resolved to ban Titebond III from the shop.

Next, we discussed the use of a shaper cutter that made shallow finger-joint-like grooves in the parts shaped with it. When clamping two boards together, the grooves effectively doubled the glue surface, increasing the panel's resistance to the elements. However, the grooves would show in the panel raises and present an unsightly zig-zag on those raises. The cutter was also expensive, and we identified a few set-up problems that decreased its attractiveness. This was enough to rule it out.

We determined that epoxy was the best next candidate. The grab-anything-and-everything-and-never-let-go grip was good. Unaffected by normal heat or cold, the dark glue line was our only consideration. But then these doors all had dark finishes already. We adopted this as process.

But then we determined that the process had changed, but it had not improved. Constant improvement was an unwritten byword (byphrase?) in the shop, a key part of process. And this change did not improve anything much. Well, the panels were not failing, so that could be called a gain.

Over the years, one element of our doors had bothered me slightly. In gluing for width, common for most of our panels, boards of solid wood were jointed together, clamped, cured, and then sanded. Often, the color of the boards or grain, or both, could not be matched as well as we liked. This detracted from the overall appearance of the doors, though we had never gotten a complaint. But they say you are the last to know.

A solution of a kind presented itself with the use of some thick (0.060 in.) veneer from a furniture project. For several years, off and on, I have played with the idea of a stable panel, one that would not pull away from the rails or stiles with the normal seasonal movement. This also was not a complaint. At least with anyone but me. I tried some three-ply assemblies. Solid faces 3/4 in. thick, glued for width, then glued to either side of a stable plywood panel. This had worked in the past, but I did not know how it had gone long-term.

Thinking about the thick veneer, I was reminded of the old cross-banded door rails and panels from 100 years ago. A cross band ran 90° to the face, and then it was bonded to solid wood. Today, the center of the assembly is a stable panel—medium-density fiberboard (MDF) or plywood. When raised, the face and cross bands are cut away, and the solid wood is exposed and receives the raise. The stable panel works in concert with the cross bands and eliminates seasonal movement—an unstated goal. All was in harmony—a good solution.

Serendipity is not a frequent enough visitor in the shop, but when it does arrive, it is all for good. While the core problem—panels cracking—was solved, another unrelated quality was much enhanced. The serendipitous result of this radical shift in panels was that it allowed us to match color and grain on our panels, allowed us to book-match the grain on all our panels, and allowed the grain to flow from one panel to the next one above or below. And, if that is not enough, we could also match the grain from face to back on a door. A great solution.

This took our product to a new plateau. Few other makers use veneered panels, and none that we know of use stable panels. No one in our market builds anything quite like it, so we remain unique within our niche.

Can any such elegant solution have drawbacks? Well, yes. Expense. The veneers and the labor add to the expense of an already expensive product. Labor is also significantly increased. In our small shop, dragging out the vacuum press interrupts almost all other operations.

A press run takes from noon one day to mid-morning the next day—if it can all be done in one run. This is all a significant disruption to our process, adding to labor and materials costs, calendar time, and more.

Confidence reminds us that process will settle in and help us get used to what the new expectations should be. We already have found ways to dovetail the vacuum operation with other shop activity so there is less dead time. Process is preserved, and has again helped solve problems, establish a new status quo, and significantly improve the product.

Chapter Three: Stock Selection, Face & Edge, Accurate S4S, & Layout

Stock Selection

Many trees will make lumber, and if you can make lumber, you can make doors. Whether those doors will stay flat and true is a question that can be answered by looking at the other local uses for the wood. Any area will have its favorite woods for various uses. Some are good for building, some for boxes, some only for firewood. Sourcing good lumber has been a long story for the woodworker looking to make doors. We typically want the best, and all of it. While others are content cutting out knots and moving on, the architectural guys will reject about any board with a defect. This limits the sources and the species that are available, and raises the costs of such wood. Both supply and demand create pressure on a species.

While almost all woods can make serviceable interior doors, not all woods will make good exterior doors. We are just now seeing fewer and fewer pine doors, but for over 50 years, pine was the most prevalent species not only for doors, but for all interior millwork. Plentiful in supply, lightweight, stable, and shippable, pine was the choice. The actual species were Eastern white pine on the East Coast for regional makers, and the Western pines—sugar pine and ponderosa pine for the rest of the U.S.

The Western pines were harvested for crating and shipping during WWII, and mill men loved to work with the wood. Light and fragrant, wide and clear, easy sawing, easy sanding, pine became the choice for about anything in the home that was made of wood. My first professional position was in a shop that used pine for 90% of its products. During my time there, the price of pine increased enough for the shop to replace pine with poplar. The local hardwood and the Indiana State Tree, poplar was coming in wide and long and clear—and inexpensive.

It was harder and heavier, but it was stable and milled well.

It was years later, as that shop was closing, I realized poplar was not a good wood for exterior work. A couple of doors had split panels on the exterior face, other than on the glue line. Then a job of shutters sprouted mushrooms after two years in service.

Stung by this newish problem with poplar, the search went out for the best door woods and what made them so. Teak was an obvious choice. So was Honduras mahogany. Then white oak, and pine and walnut could be added to the list. Douglas fir was also added on the softwood side. The imported woods tended to shift around in availability, and the question of sustainability was also raised. Field performance was not assured since these species could not be observed like pine, oak, and the others could be. Poplar, alder, red oak, cherry, maple, ash, and elm all made the undesirable list due to direct experience, hearsay, or lack of experience.

The prevalent thought has been that a good exterior door wood was the result of species. Certain species are sufficiently able to repel water or fungi or bacteria. Cedar and redwood have a natural repellent to insects and bacteria. The better woods stay in place after exposure to high moisture, low moisture, high heat, and ultraviolet extremes. But there are a lot of broad interpretations of what makes good wood or a failed wood. In hearing these, I can never be sure: Has the describer used a scientific approach? Is this all hearsay? Does this guy even know what he is talking about?

There is good science on wood as a building material, and lots of research on what cutting angle is best for a Forstner bit in Eastern pine, but no real science on what is termed "durability in service."

A large state agricultural university researcher recommended poplar as the best wood for a window manufacturing startup. He made his notes and started up a company based upon a local hardwood. After a few years, the poplar windows started to grow fungus or rot, or move far out of line. The window-maker was inundated with warranty claims and angry consumers. A good effort was made to remedy the situation, but with the company rooted in poplar as its only wood, they were forced to close.

In retrospect, the state agricultural university explained that poplar was the chief wood selected by settlers in the Midwest, as they built cabins from the wood felled when clearing land for farming. These poplar logs cited by the university were in log cabins at a nearby pioneer center and historic recreation of a small town in 1836. These cabin logs, cut as long as 175 years ago, were still solid, with little or no trace of rot or fungus.

We are now of the belief that ring density plays a major part in the consideration of a wood's suitability for exterior service. The annual rings present in the cut surface of a log record its growth, and will vary greatly in their density. Some old-growth fir and redwood will have well over 150 rings—or years—per inch. Current poplar in the shop has about seven rings per inch. The old-growth woods are not only extremely stable but are also able to repel most of the bugs that like to eat the wood or its components. The poplar logs (and other species that were equally old) at the pioneer center had ring densities that are nearly impossible to count, and all in dark, dense heartwood. While a good observation, the fact is not a scientific approach to examining the durability of the species.

This fact reduces the list of best exterior door woods considerably. When buying lumber, one cannot stipulate "100-years-per-inch density" or similar. One can ask for, and sometimes find, old-growth woods, but it is not something to count on. And do we really want to cut down those magnificent giants to make some doors?

It is worthwhile to recall that lumber—trees—were literally in the way of development in this nation's early days. Pioneers settled on land and cleared it so as to "improve" it and get title to it. In the Midwest, large tracts of trees were girdled so as to kill them, so crops could be planted between them. Once these trees had died and dried on the stump, they would be set afire to get them out of the farmer's way. Some nights, thousands of acres would be ablaze as these virgin tracts of timber burned for days or even weeks. Billions of board feet of the most beautiful lumber in the world…all gone.

Similar scenes are still played out today, all over the world, as the great forests are cut for lumber, cleared for agriculture, and even stolen for profit. Sustainability has come into play during Acorn's existence. I cannot source only old-growth giants of the forest for my product. I must do my best with what we have, and respect that and all that is implied by forestry industry best practice. With the population growing, arable land shrinking, and forests being felled everywhere for everything from living space to agriculture, to chopsticks, sustainability is now more important than ever.

Sustainability has played a role before. It is said Rome lost its grip on the Mediterranean as its forests were consumed, first for the fleets of ships required to police the Roman Empire, then later to the making of iron, which required large amounts of charcoal.

Later, similar conditions pressed upon Great Britain as it worked hard to build and maintain a large fleet of ships with which it ruled its empire. Indeed, large tracts of forest in the new land of America were marked "Kingswood" as the property of the king, and off limits to any colonists. Eastern white pine was earmarked for masts and spars, while live oaks in the South were prized for the heavy branches that had a natural curve along the grain and therefore could support the ever larger guns that the British were using. Teak and mahogany were planted in Malaysia and Indonesia to help insure a supply for the future.

Increasingly, Honduras mahogany (*Swietenia macrophylla*) kept surfacing as the best choice for doors on all points. Today, the harvest is protected, as trade in Honduras mahogany is now regulated by the CITES treaty and policed by satellite. The wood is plentiful, though scattered throughout Central and South America. It grows singly, rather than in groves, so it is costly to harvest, with helicopters now being the preferred method since it avoids the building of a road and the settlement that follows road-building in that part of the world.

The stability of the wood is displayed by one of the sub-grades under which Honduras mahogany is sold. "Pattern grade" Honduras mahogany was segregated out for the pattern-makers of the U.S., for their demanding patterns that had to be stable and repeatable. Today, the pattern grade is still used by pattern-makers, but also for door-makers who like its smooth, even grain and texture, with a limited range of color and density.

In working with the wood for 50 years, I have seen the size of the boards get shorter and narrower, but the wood is still the best door wood. It even has the added cachet that it is "mahogany," a prized and expensive wood. Unfortunately, good quality custom exterior door production relies upon Honduras mahogany instead of a cheaper, more readily available wood—like poplar. Poplar remains a good wood for interior use when properly dried and conditioned, but is has no place in exterior use.

Selecting the species to use may be up the to the maker or the owner or designer. When odd woods are spec-ced, politely ask why that wood was selected, and be ready to make your recommendations when you realize their selection was arbitrary or misinformed. A warranty can go to work for the maker at this point when the designer realizes blue-berried babawaka is not covered by the company warranty. Use your knowledge to help select the best wood for the project. But, like Henry Ford, be prepared to just say black—or Honduras mahogany. Although, black walnut, white oak (flat, rift, or quartered), white pine (sugar or ponderosa, clear or knotty), and Douglas fir are all excellent domestic woods for doors. Western red cedar is a good exterior wood, but is thought to be too soft for passage doors. Teak is also a fine door wood, but it is so expensive for larger cuttings (about 3.5 times the cost of mahogany) that it is almost never used.

Examine your rough lumber by placing enough to make your door on the bench or horses so you can flip the boards over and see all sides. Find the defects and mark them. Look for the large parts first—the stiles. They need to be clear (assuming the door is not supposed to be knotty) and straight. At least straight enough to be "straightened" on the joiner and still make thickness. This is usually a source of trepidation for any woodworker. It is an acquired skill to be able to confidently

bust out big sticks that will indeed clean up at the planer and make good door stiles.

"Busting out" is the term we use for filling a cut

list with parts that will make the final sizes required. The cut list is a clean, organized list of parts notated in their final size. Every job will require a cut list, so plan on making one every time, and using it. Make your entries and notes the same way every time, and the list will stay organized and help you get the parts made for your project. Walk away and come back a day later, and you will know where you left off. A convention in every shop I have worked in is writing the three dimensions thusly: thickness x width x length, or 1¾ in. by 6⅛ in. by 97 in. Once you have a clear cut list, you can start cutting the real wood.

Well, maybe cutting the real wood. If you are unsure—your skills, so much wood, expensive wood got you all bound up—use a pencil first. You won't be the first to experience a stall at the saw table. Mark out on every board what you plan to yield, and go through the boards until you have all your parts marked on the boards. Move the pencil marks as needed, scrape them off and remark them as needed—no problem. All woodworkers have experienced a moment or two where we were paralyzed before the board. We just could not gather our wits enough to cut it. With experience, this will diminish. If you develop a clear plan and follow it, your confidence will grow, and you will develop a solid foundation. You need both.

We cut for length first, finding the longer parts first. These are cut at one end to get rid of end checks, if any, then at a rough cutline, usually a couple of inches longer than called for on the cut list. Next, the board is rough sawn to width on the table saw. In the Acorn shop, all parts are roughed out for length on sawhorses with a little cordless DeWalt circular saw. In previous shops, rough lumber cut for rough length was almost always done on a radial arm saw.

A 10-in. Delta Unisaw, 3 hp, with an accurate digital readout is put to good use rough ripping the parts. Ripping over width is minimal—usually 1/4 in. to 3/8 in. over final size. Leave enough for straightening on the joiner as part of facing and edging. If the board has a bow, you may need to increase the rough rip width in order to leave enough to be cleaned at the joiner for straight.

The joiner has become an odd machine in many shops. However, for the person charged with making things straight and square, standing on its own, the joiner is invaluable. Simply said, it makes the wood straight and square on two adjacent faces—a face and edge—while cleaning up a surface well enough to use as a flat reference when run through the planer, table saw, shaper, or other machine.

Face and Edge

The task is to plane a straight—or flat—surface on the board. No twist, no curve, no slant, no shift or wiggle. A board is placed face down on the joiner's in-feed table, with the depth set about 1/32 in. or less. A facing pass is made, holding the board down on the in-feed table, but not pressing so hard as to deform the board from its natural state or to cause difficulty in advancing towards the cutter. Pass it over the cutter, with the left hand pushing down lightly just after the cutter, and continue the pass with the left hand taking the advancement pressure, holding the freshly cut face to the table, the right hand pushing until the tail is out of the cut. Listen carefully to the sound of the cut. It will grow louder and softer, and the cutting resistance will change slightly, as more or less wood is cut away.

This is why hearing protection is worn at the joiner—to be able to hear, even after 50 years at the machine. Do not hook the fat of the hand or fingers, or anything, over the head or tail end of the board, as these will be trimmed in no time. Push blocks should be in a handy location for use when needed. Make or buy the kind that have an adequately sized handle with a small hook on the underside that can be used to push those large and heavy boards. Smooth and fluid, the act of joining or facing a board will become easier with time. It is a valuable skill and a crucial step in the building of doors. Enough so that, if you do not have a joiner—or some foolproof method of making things flat—then your time will be better spent on something else.

Another way of jointing is to start with a piece of wood the length of the in-feed table. Place the less-than-straight edge down to be straightened. Secure the piece down with the left hand, just as dead weight, to keep the piece from bouncing around, then advance the piece forward with the right index finger. Finger only. Just let the board find its way through the cut. Once the entire board is on the out-feed, out of the cut, give it the eye. It should be dead straight. If not, this is the point to remedy the skill. Keep practicing until you can do this without much thought. Just let the board find its way over the cutter, no force, no brutality. Smooth, flowing effort that taxes the body less and gives better results.

Jointing is an activity that can take all day to do hundreds of parts, or just a few minutes out of a day. No matter, it requires attention to do it satisfactorily, as we gradually learn that lumber will make a better thickness than what it seems when facing. This gives us confidence, and helps us become better at this skill. Once acquired, it is like riding a bike. You never forget, and you can do it with ease and maybe even a bit of style. But honestly, I would be very wary of getting on a bike, not having ridden one for about 30 years. But joint a board? No problem.

Jointers—the machines—come in all sizes, colors, and types, but they all perform that one simple function: They impose a flat surface, a plane, onto a board. There is some confusion over the two terms—joiner and jointer. I feel that a joiner (person) uses a jointer (machine) to make joints (noun) by joining (verb) pieces of wood—all the correct ways to use the language. Accurate communication is what counts, and as long as everyone in the conversation understands, you can call it what you like.

The old gentlemen I learned so much from 50 years ago never had any formal training in woodwork, just years and years in the shops. Their terminology and logic were seriously lagging, as no one could verbally explain how to joint a board. Give them a board, and stand back and watch, and you could easily learn more than you could ever ask. Emulating their actions taught me enough to get me going, and to eventually gain confidence in my ability to straighten a board. I learned the process from an 82-year-old man who had only one eye, and a total of about 8½ fingers, but he could joint lumber without a thought. And do it safely.

Jointers come in sizes from 4-in. desktop model-maker joiners on up to 12-ft.-long, 18-in.-wide machines that resemble aircraft carriers. Ship-builders used 24-in.-wide "facing planers" to surface and shape the parts of a ship, and their equipment included adjustments for draft or taper, and for making hollow joints. New jointers today, with one known exception, do not offer such advances. Northfield Patternmakers Joiners are notable for their ability to tilt the tables so as to run tapers and such. The other notable characteristic of the joiner is the ship's wheel depth adjustment. A spin of the wheel adjusts the table about 0.005 for each full revolution. Hair splitting capability.

Depth of cut is adjustable through a large hand-wheel, small hand-wheel, or lever. Depth of cut is from zero to as much as 3/8 in. The deeper cuts are for edges, and are too dangerous for wide surfaces. A 4-in. board would typically be set at just under 1/32 in. for a facing cut. Use the planer if you wish to remove wood. Short lengths—less than 12 in.—should not be jointed. Edges can catch and flip the boards, or shoot them out from under your hands. When jointing, it is advisable to raise one's hands off the board as it passes over the cutter, as a just-in-case measure.

The joiner—and shaper—are two machines that can cause a lot of damage if misused. While a band saw or table saw can cut off a part that is sewn back on later, joiners and shapers make hamburger. It was once thought that, in order to work wood, a body part must be sacrificed. One's nickname may become "Lefty" to denote what is left or missing. This is, of course, not true, and it should be expected that a person can work a lifetime without serious injury. I still have all my parts, with only one corner of an index finger, trimmed on a non-electric miter chopper.

A long-held safety consideration is to think: "What would happen if that board disappeared—where would my hands go?" This is why we lift a hand as the wood passes over the cutter head on the joiner. We are often pushing wood towards a cutter of some sort, so it is worthwhile to pose such questions. "What if…?" A band saw can just as easily cut off a thumb as cut a board in half. Asking yourself that question, "Where would my hands go?" is often enough to create a good proactive mindset that will keep you counting to 10 up into your old age.

If fear is your only, overriding response, then back away, and get some more training. Education, then experience, will bolster your ability to turn on the machine and make productive shavings.

I think it is notable that confidence plays a role in being a successful woodworker. I am sure it plays a role in other endeavors, but I think it is unique in woodworking. First, there is the lumber, which intimidates and must be overcome—tamed, in fact. Then there is the equipment. Equipment that does not care who you are or where your hand is. It does what it does; it is our job to learn to stay out of harm's way. Even to go out and sell yourself as a craftsman takes confidence. After a few years, your confidence builds, and you are more comfortable in the daily jobs about the shop. A sizable setback of some sort can shock your confidence level and weaken it temporarily. Go slow, lick your wounds, take some time, then get back in the saddle. At worst, you may require listening to Stuart Little's affirmations.

Mass matters. A 60-lb. 4x4 cedar post fed into a 1/32-in. cut on the jointer will be a light, easy pass with even pressure. A 2-in.-wide, 3/4-in.-thick, 2-ft.-long board, while the same cut, will feel dangerous, even with push blocks. This is because of the ratio of depth of cut versus the mass of the wood. The heavier the wood, the larger the cut that can be taken. The feel of that cut is learned, but not instantly. Shifting ratios of size and mass and cutting pressure move around in the head as the board is fed into the machine. Trust your developing instincts here, and back off the depth of cut to 0.015 in. or so, and it will feel safer, and give you a better joint. The machine is not a finishing machine for the most part. The surface it creates can be smooth and ready to sand, but beyond face planing, it is used mostly for sizing large panels to width, or for fitting cabinet doors.

Cutter heads for joiners are now all round, with narrow slots for the knives and the locking gibs and bolts. Three or four knives will be fixed into the head, adjusted so they are all at the same cutting circle, and rotated at 4,000 to 6,000 rpm. High speed steel (HSS) is the usual knife material, with carbide available. Some joiners have Tersa heads that use disposable knives, and change in a few minutes or less. These have the advantage of quick-change design; new knives are merely slid into the voids, and the machine turned on. Centrifugal force pushes them into place and the machine is ready to serve.

A recent advance in cutter head technology uses segmented carbide "chips" in place of steel knives. The head holds the chips in a spiral arrangement that presents the cutting edge to the wood at an angle, reducing the cutting angle of attack, making the cut slightly easier on the equipment. All the chips work together to make for a quieter cut with less tear-out. However, carbide does not give the same quality of surface that HSS does, though it is still very acceptable for glue joints, etc. One clear advantage of segmented cutter heads is that a lower noise level is produced. One noted drawback is that the carbide chips can create match lines on the wood, where misalignment, or some variable, is introduced to the carbide chips, causing them to create a less-than-perfect surface. This may be a quality issue in the cutter head, or it may be part of the process that is unavoidable—it is the natural effect of the many chips. Remember, the joiner's job is not to provide a finished surface. Rather it is to produce a straight, flat face and edge, and the two surfaces become reference points for the entire board.

Once the face is flat and clean enough to be a reliable reference for planing, then take the board, tip it up on its long edge, and pass the edge of the board on the joiner, with the freshly cut face snugly sliding along the fence. The fence must be a reliable and sturdy 90° to the table. Keep a rigid square nearby for setting the fence and verifying its angle during use. It may take several passes to find the straight edge you seek. Adjust for a deeper cut if you need to. Shift the hold-down pressure from the in-feed table to the out-feed table, listening to the cut. When simply looking at rough hardwoods, you will notice that the edges deviate from straight far more than the faces of the boards. At times, lumber can appear so squirrelly that it is a wonder that anything can be built from it. However, with practice, it will become possible to use even wavy and bowed lumber if it is thick enough to take several passes.

The board will now have two straight surfaces, square to each other, and dead straight. Emphasis on straight. If the board is not straight, run it again. Depending on what you are working with, facing may take one to three passes, and the same with edging. Pick an eye, and learn to eyeball the wood. I use my left eye, and I have sighted many thousands of boards. Your eye will get faster and better with practice. In our shop, everything gets processed on the joiner, so it will all stand straight and stay straight. Upon occasion, a board may move after joining. This is due to reaction wood in most cases. Reaction wood has tension in it, and is often first seen at the rip saw during rough width bust out.

One reason the joiner is the odd machine today is that, while its function is clear and understandable, how to get there is not a direct process. It takes some experience and some confidence, a bit of luck, and few losers before it gets to be a part of you. An element of surprise is included since one does not see if the stock cleans up until it comes out of the planer on a final pass. Nothing like a little suspense in the shop.... As long as we remain firmly in control.

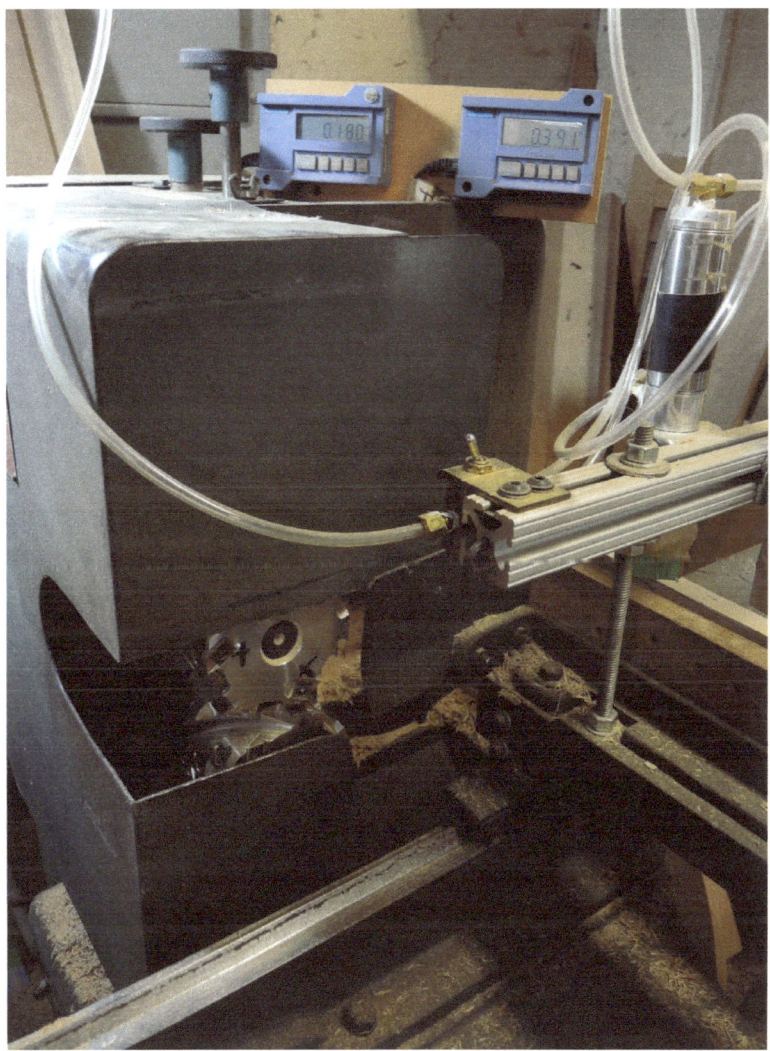

Accurate S4S

Accurate S4S, or "surface four sides," is fundamental to fine woodwork of any kind. The next step to get us there is the planer. This machine will bring our parts all to the same thickness, referenced by the faces made previously on the joiner. Smaller planers will only open to 5 in. or 6 in. maximum, so often door parts cannot be "edged" through the planer. They are then joined clean on the original edge, and then ripped to the number, or ripped oversize. Ripping is important in that both edges need to be parallel and square to each other. While not as accurate as a machined surface, a ripped edge can be very true and adequate after sanding.

A planer suitable for small shop door production should be a few horsepower, have adjustable in-feed rollers, and enough power to run for hours at a time, making a 1/16-in. to 1/4-in.-deep cut all along. Dust collection needs to be robust to carry away the shavings and not clog the machine. "Lunch box" planers will not work hard all day, so are not suitable for daily door production. As you can see, equipment suitable for door production is easily heavier and more robust than cabinet shop equipment.

Lumber is fed into the machine, taking a pass that is not too heavy. Plan on hitting your target thickness within three to four passes. Send all the boards through the machine with the jointed surfaces down so the planer can use the flat, jointed surface to reference its cut on the upper surface of the board. If your machine has adjustable feed rollers, these can be set so as to allow the wood to break friction with the table and feed more easily. After the first pass, each board can be judged—did the planed surface clean up—get cut—completely? Next pass, etc.

A recommended accessory for any planer is a digital readout or DRO. This device will accurately measure the cutting dimension to hundredths of an inch or even thousandths. You can see how valuable this is when making parts. With a good DRO, one can change the planer thickness setting and yet go right back to the exact setting by means of the DRO. Acorn's shop has accurate digital readouts.

Accuracy is another value in the shop that has some variability. How accurate is accurate? Acorn has repeatedly learned the values of increasing accuracy over the years. We have added DROs to the tenoner, mortise, shapers, saws, and planer. These reduce or eliminate set-up times, allow for repeatable settings, and help make precise, controlled adjustments and accurate S4S with a high degree of predictability.

Paired with a good digital caliper, settings become more rigid and less "squishy." We can adjust the fit of a tenon from hammer tight to rattling in the mortise. And do it all in a matter of a few moments with the DROs.

Planing a batch of parts can be drudgery. It helps if there is a variable speed feed on the planer that can be sped up on the first passes, and slowed down for the last passes. It also helps to have a helper catch the parts and stack them for the next pass. Generally, it is the off-bearer's job to turn the parts so the next surface to be planed is up, freeing up the in-feed person to see what is to be planed next. Be sure to stack the parts neatly and logically, on a sawhorse or a cart, so they can't fall or become unstable.

Now we can look over our nice stack of door parts all freshly milled and sized, ready for the next steps. If any gluing for width needs still to be done, now is a good time to do it. Wood that is freshly cut bonds better than older surfaces—older by as little as a day or so. The joint is not at risk of falling apart, but it is not as strong as it could be. As it "ages," the surface oxidizes and will not accept glues the same way as if it were non-oxidized. This is an important characteristic to note when running large numbers of parts that get joined later. Yet another complicator.

Put your stiles on the bench, crown them, and mark out the mortises, haunches, and sticking as needed. Mark your rails to length, then mark the tenons.

Layout

Layout, or "setting out," as my elder mentor called it, involves marking the wood for the joinery that will follow. Start with the stiles. Crown each stile by using the eyeball. Mark the high side, as the high edge, of each stile with a "V." Pair the stiles so the crowns are up and opposed to each other. If you are good, and your lumber is good, you will see there is almost no crown in your parts. That is fine—you can mark either face as you like. If your stiles have any crown, place and mark these opposite each other so they do not touch in the middle, but do so at the ends. This gap is twice the crown, if both crowns are equal.

Crowning a door was once common. Today, it is rarely heard of.

You may become a door builder who likes to crown your doors by intentionally putting a 1/8-in. crown (or more, or less) into the plane of thickness of your stiles. This is an advanced technique on the joiner, for the advanced characteristic of a crowned door that snaps shut, with tension on the latch bolt, and tightness against the weather seals top and bottom. Old timers, if there are any left—liked to crown doors. Most younger carpenters have never heard of the practice, but the good ones will unintentionally work with the crown when they sight their eye down one edge or the other, then machine the door to work with the crown.

Mark the edges of the stiles for the actual landing place of each rail edge. The stiles should be one or more inches longer than needed. Use a sharp pencil, and dent the wood slightly at the mark. Then use a square to extend this mark across the width of the paired stiles. Or all stiles if they all get the same rails/joint. Repeat at every point—rail profile, rail flat, end of rail, end of mortise, end of haunch. Mark both stiles (or four or six or eight...) at the same time for accuracy. This is also a good time to measure twice, checking against paperwork to verify you are still on the right path.

Layout the rails with a square, marking two lines, one at each end for net length. A formality acquired almost 50 years ago has been "circle numbers"—the number in a circle is the length of the rail between tenons, shoulder to shoulder. The length with the tenons is a "square number"—the overall length—and is written in a square on the part. This comes from the cut list that lists every part, quantity, size, and details.

Some rails will have muntins intersecting them. Now is a good time to locate where the two parts will meet. Accuracy in locating the muntins is important, since they should be glued in place.

Now is a good time to make a "face mark" on the parts other than the stiles, which already received their face marks at first layout. This is just a handy single pencil stroke on one face of all the individual parts of the door. "Face marks up!" is the mantra to hear in your head as you mark your parts, and then when you assemble. The face mark will be "up" in the tenoner, the mortiser, the shaper, and at assembly. This will make sure all the parts are going the correct way if anything is intentionally or unintentionally off-center. Almost all our doors are symmetrical about the center. The mortise is centered in the thickness, the panel is centered in the plow, which is also centered, and so on.

Despite our best efforts, and a fine pair of digital calipers, it is not unusual to be a few thousandths off. The face mark will put all the "off-centeredness" on one side of the door, minimizing that few thousandths. Were we to put a rail "up" at assembly into a stile that is "down," the few thousandths would double. This will leave a bit more than we may want for later sanding. While the reversed parts should not be offset more than 10 or 20 thousandths (0.010 in. to 0.020 in.), it is not too hard to sand flat. However, the essence of craft is that little trick to get everything level, no offsets. One more thing dealt with in a manner that will promote quality instead of detract from quality.

Often, which "side" is the face does not matter, but it is still a convenient management tool to help keep everything organized. Faced with a door assembly of 20 parts, face marks remove at least one question: "Which side up?" Typically, a door is assembled on the bench with all face marks up, and looking at the "back" of the door. Or the interior side if an exterior door.

The face marks on the stiles, marked when the stiles are first set on the bench for layout, now denote the interior side of the door, with that side facing up. If there is a crown to be preserved, both stiles will have their crown up. The opposite side of the door will be to the weather, and also to the stops in the frame. While one can build doors without face marks, face marks certainly do make it easier as complexity increases. A four-piece, open light door may not need the level of organization that face marks provide, but most projects are more complex, and their use becomes evident.

The stiles are now all marked out, and set on their "back" edge, with the mortise layouts up and visible. Rails are stacked so one can visually find the top, mid, and bottom rails easily and logically. Face marks all facing the same direction, preferably up. Once glue is in the mortises, and the rails and panels are going in, things tense up, and it is nice to be able to tell which face of any part goes in which direction. It is time to move along.

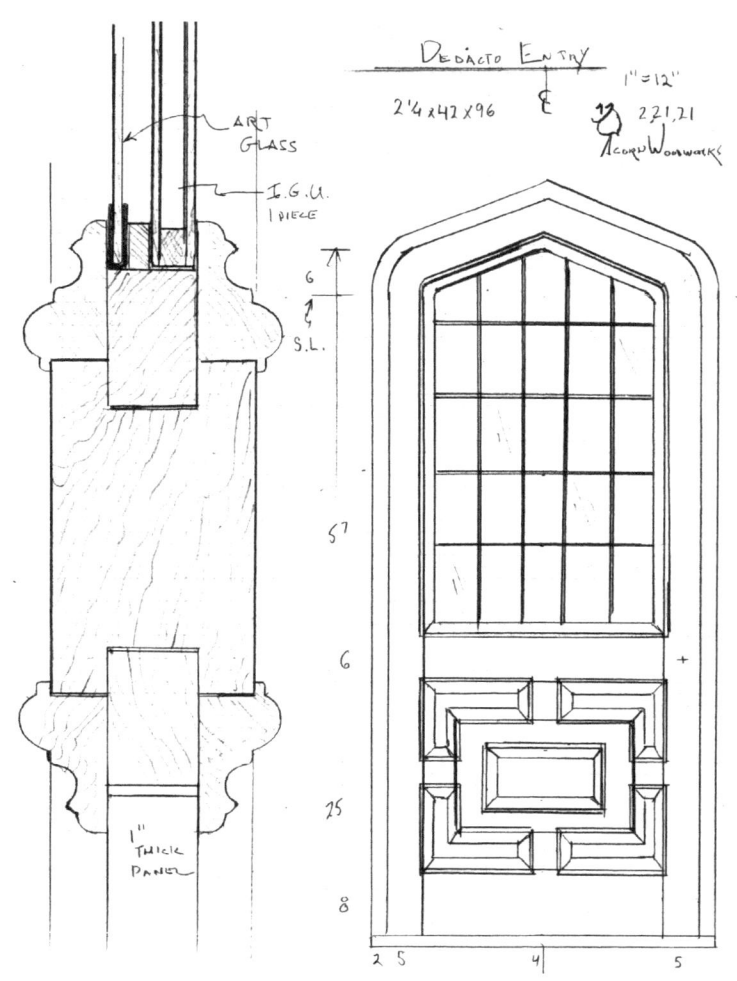

Chapter Four: Joinery, Tooling, Hand Tools, & Glues

Joinery

Henry Ford was said to offer his Model T in "any color, as long as it is black." This simplified his assembly lines and helped revolutionize the industrial revolution with mass production. Acorn has much the same feeling about joinery. Any joint, as long as it is mortise and tenon, were anyone to ask. Given the low level of knowledge the buying public now has about wood products, most people will know nothing about the best joinery for a door. If they learn a little bit, they can easily see why it is the best, with no other clear options for good build quality.

Every woodworker knows two things about mortise and tenon, both of which are mostly true. The joint is hard to make, and it is the best. "Best" meaning strongest, least likely to fail. The joint is designed for joining right angles or near right angles and, as such, is used in chairs, timber frames, and any place a frame is needed. Chairs found in Egyptian tombs were made of ebony, and had mortise and tenon joints that were still tight after 6,000 years.

The argument can be made that working wood is the oldest profession. One tribal member knew what wood made the better arrows or spears or musical instruments, and they were called upon for finding and utilizing those trees, those woods. This was a natural extension of knowledge gained from collecting herbal treatments, as well as foods, both common and rare.

Certain woods were prized for various qualities valuable to the tribal community. Indeed, whole trees of exceptional size or location would be revered by our ancestors. Most recently, we have "big tree" lists that help protect the trees we value as special. Not long ago, special trees became known as meeting places. These were favorite places to sign treaties with Indigenous people.

The difficulty in making the basic joint lends itself as the foundation for some sort of philosophy: "Nothing easy is worthwhile," or similar. Perhaps woodwork—and the mortise and tenon in particular—is the foundational, solid structure behind the philosophical component present in so many beliefs, that the path to enlightenment or other lofty goal is not to be tread upon lightly or easily. If pain equals success, then that might explain why so many woodworkers are smiling. As professional woodworkers, we certainly know pain.

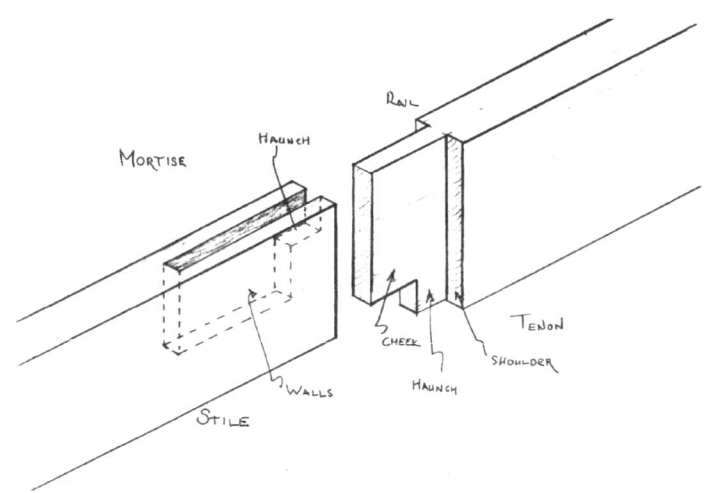

The word tenon has been in use since the 14th century with about the same meaning. Tenons can be square, rectangular, or cylindrical in shape. In door work, almost all tenons are rectangular. The tenon shoulders fit to the edge of the stiles, while the tenon cheeks fit into the mortise and give the bulk of glue surface contact.

The point or function of a mortise and tenon joint is to increase glue and/or mechanical contact surface. Butting the end of a 2x4 to another 2x4 side grain, with almost any glue, will result in a weak joint. The 5¼ square inches of contact is all the contact there is. The joint also is not protected—not enclosed or sheltered from abuse or weather. It will not last long.

If we were to make a 1½-in.-long tenon on the end of a 2x4, and fit it into a 1½-in.+ deep mortise, we will see the glue surface change to 14 in.—nearly three times the wood surface. The flat surfaces that fit together will do well at resisting racking or twisting forces. Plenty of surface for the glue to bond to, giving it the racking strength required to make a solid frame.

Mortise is also an old word with English roots dating back to the 14th century. Some will date it as even older, sharing the root words for *mortare*, to die, and *mortuary*, a place for the dead. The connection is the fact of the mortise's shape—like a grave. Again, some will also trace the root to the word for mortgage—or to pay until death. Not the most uplifting, but it is only a word.

The mortise is clearly the more difficult of the two to make. While a good bit of handwork, excavating a mortise is obviously worthwhile since we have seen it used for thousands of years. When looking at preserved old structures like cathedrals, Asian temples, fortresses, bridges, and even simple houses, we can find the mortise and tenon in use wherever wood is used as a primary building material. The joint has crossed oceans, traversed mountains, and traveled to every corner of the world, as people use the joint for their most durable applications. Most cultures have used it for so long that its origins are lost in time.

Today, the average woodworker stands at their bench and is intimidated by the mortise and tenon. Not because of its long and laudable history, but because it has humbled every woodworker on the planet. It will humble us today and tomorrow also. Anything so useful, so basic to the success of joining wood at right angles should give one pause. Respect is to be paid, as they say.

The pause will give us some time to consider the forms the joint can take. The simple mortise and tenon is as described above—rectangular sections. This is by far the most common form of the joint, used in timber framing of all sorts. The next most common mortise and tenon joint could be the round tenon, as commonly found in chairs of many types. The round mortise greatly simplified mortise-making since it was a hole that could be bored by various means—all of which were easier and more accessible to the woodworker of the day.

Millions of chairs have been produced with round tenons as the principal joinery. The Windsor chair is a tour de force of round joinery, with tapered tenons for the leg/seat joint, and many smaller tenons for the back dowels to fit into the crest rails. In the Windsor, properly done, the joints hold well enough to let the chair have some flexibility as it conforms to the seated individual.

The basic mortise and tenon joint can be reworked in any number of ways to further utilize the unique properties of wood. The joints can be doubled, interlocked, housed, haunched, through, drawn, wedged, and fox-wedged. Just for starters.

For instance, the double mortise and tenon effectively doubles the glue surface area. These can be used when the cross section of wood is low, but the demands on the joint are high. Driveway gates will get double mortise and tenons since they have to reach eight feet or more, and do so with smaller frame members, all while remaining perfectly square. Leverage is not such a factor in a normally proportioned passage door, so single mortise and tenon joints will suffice.

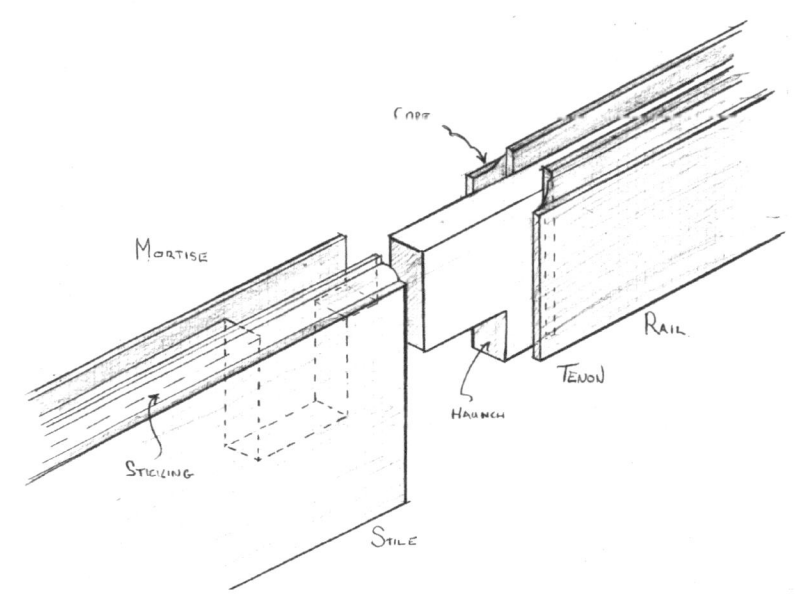

Another way of increasing glue surface in a mortise and tenon joint is to add sticking—or profiling—on the rails and stiles. The end of the rails, at the tenon shoulders, has the profile of the stiles cut into the end of the rail in reverse so it is a perfect mate to the profile. Then as this joint is brought together, the extra glue surface is seen in the cope and stick joint.

Going back to our sample, we last had 14 square inches of surface area in the joint. If we add a common cope and stick—ovolo—we find that another 8.4 square inches can be added to our joint, for a total of 18.9 square inches of glue surface. The butt joint was only 5.4 square inches of contact.

One other cope and stick joint that is often used, but not always glued is the muntin bar coped joint. Muntins are the narrow pieces that divide the glass into so many lights. In many Acorn doors, the muntin is only 1/4 in. wide, with 5/16 in. to 11/16 in. sticking out on either side. The muntin bars fit together to make the divided lights, and may have tenoned ends to fit into the stiles (and be glued), or just coped ends to fit up to longer muntins to make the divisions. When the muntin bars are in the assembled door, they are often loose—or at least have no glue. The sealant used to seal the insulated glass units in place works with the glass stops to hold our muntins securely in place and prevent any movement while aligning all the sticks correctly.

While the cope and stick joint is prized for its superior glue surface, it is more difficult to make than the basic mortise and tenon. No Free Lunch. Cutting in end grain—coping—is not easy. Fortunately, modern machines, cutters, and clever woodworkers can make it happen. Safety is still paramount, and understanding the forces at work when end-grain coping is important. Large diameter cutting discs can have a lot of wood "in the cut" and, if allowed, can throw that wood out with surprising force. I have heard all the stories, most of which I discount as folklore. However, the folklore has an underlying truth—these machines can hurt a person. Guard the cutters well with a sturdy build, thinking all along, "What if this thing gets loose?" Coping can be repetitive, with many parts, so be sure to guard against the inattention that can come with repetition. Fortunately, in a small shop there are many things needing attention, and they will provide a good break from any dulling monotony.

Other types of joinery used in our doors: the venerable panel and plow, the miter, stack lamination, and rabbeted butt joints in the frames. The panel and plow is simply the groove made for the panels to fit into. The groove can be made to the fit the panel, or the panels made to fit the groove, whichever is best. If stops are made, then the panel fit will be made by the loose stops.

This is probably a good time to mention how Acorn panel plows work. The plow is arranged so the panel plows, the glass rabbets, and the tenon thickness are all the same. Most 1¾-in. doors get 1/2-in. tenons, so raised panel edges will be "raised" to 1/2 in. thick to fit within the sticking. Doors that are 2¼ in. thick get 3/4-in. tenons, and the same for the raised panel edges fitting within that sticking. This avoids the awkward transition with panels and plows demanding one plow size, and copes and tenons demanding a different size.

We grind our own high-speed steel knives for raised panels. Three knives of each pattern fit into a 7-in.-diameter steel head. The flat that makes the tongue for the raises is ground so it is not quite flat, but has perhaps a 1° taper. This helps get the tongue started into the groove at assembly, but gets tighter as the stiles and rails are clamped together and the panel edge is driven into the plow. We aim to have that last bit of the raise very tight so as to prevent water from getting past that edge.

Some doors will still require different plow and tenon thicknesses, but by trying to minimize arbitrary sizing, transitions are easier to deal with. A good set of drawing sections will help visualize the joints and help the maker determine which is the best route to take. One does not choose to change things around from glass panels to raised or flat panels unless one has to. Simplicity, limiting set-ups, and devising a plan before the wood is cut are all important steps to be taken when faced with more than one panel plow/tenon size.

Miter joints are rarely part of the stile and rail structure of a door. They are almost always a part of glass stops. Semi-structural miters come into play with bolection moldings (a molding that projects beyond the surface of the door), which may retain panels and/or glass. Miters also play a role in the use of rimmed panels where a stable core is picture-framed with mitered stock, then veneered, then raised to make a stable raised panel. Miters can be made a variety of ways, and accuracy matters. Take the time to prove your miters.

Bolection moldings in this shop are assembled off the door, so the miters can be reinforced. A bolection molding is a molding that projects out above the face of the stile or rail as it frames the panel openings. Reinforcing a miter joint can be as thorough as a cross-grain spline fit into a groove in each half of the miter. Or it can be pinned or stapled, or even screwed. Use plenty of glue, have a wet rag ready for clean up, and reinforce those miters. Acorn uses a ledger strip centered in the thickness of the door, to seat, or "land" the bolection moldings, adding to the reinforcement.

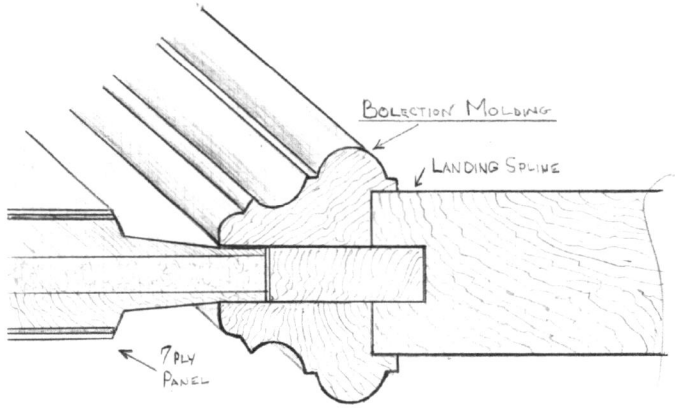

Often, lumber in the needed thickness is just not available. This is when we use stack lamination to make stiles, rails, and frame members that are of the rough thickness we need.

Lumber today is mostly graded from the best side, so for those of us faced with making anything that is visible from both sides, we have to develop a strategy to get appearance grade on both sides of a board. Stack lamination allows this. Rift white oak is a fine door wood, but comes to us a bit narrow, and with only one side attractive. So we will buy 4/4 lumber in the rough, and plane it for three-ply laminations (13/16 in. to 7/8 in. to clean up) to make a lamination ($2^{7}/_{16}$ in. to $2^{5}/_{8}$ in. thick) that will then face and edge, then plane to make 2¼-in. parts. We plane it to clean up, and then put the ugly one in the middle, and the better boards face out in the lamination. With this example of rift white oak, it is possible that we will have to glue for width, then glue for thickness. This adds considerable labor and material costs to the project. We face and edge and plane more than two times the amount we would if we had thicker lumber.

If the thicker lumber can be found, it often will cost so much more that it will be the same as cost it would be if we were laminating three boards together. Then it is our choice—spend a little more, get the thicker lumber, and process it more quickly, or get the thinner stock to laminate, and spend more time to get the parts needed. It is an old story—more time and less money, or more money and less time.

Rabbeted butt joints are about the only other joinery, and will be covered in the discussion on frame parts and assembly.

There are many variations on the venerable mortise and tenon joint that go well beyond the length of tenon and width of the joint. Coping a joint increases glue surface and adds an attractive and useful profile to the door. Cope and stick cutters are mostly custom made to each shop's directives. At this writing, cope and stick cutter sets on the shelf are merely sets that produce a "stuck" tenon—a tenon less than 3/4 in. long, formed as part of the sticking, not as a true mortise and tenon. There is no mortise, only the panel plow. This does not yield the surface area so obvious with a real tenon.

A double tenon was mentioned as a variation on the more basic form. Round tenons were also mentioned. But, for doors, the selection is narrowed to two or three joints—square-edged, coped one side, coped both sides. Thickness, width, length, haunch, and more will vary infinitely, with the basic joint being the same.

Mortise and tenon joints are often spotted with a peg that runs through the joint. This pin can be placed after assembly, where the pin is decorative as well as functional, or it can be placed during assembly. When the bores for the peg are done, they are offset in such a way that the tenon bore draws the tenon in, with the tenon bore being offset in relation to the bores in the stile. The tapered pin is driven in, pulling the joint tightly together. This is called draw boring, and can be done in 3/4-in. thick stock as well as timber frames. In timber framing, the pins are called tree nails, and are about one inch in diameter, made of green locust. I have dismantled timber frame joints that are over 140 years old, with tree nails that have permanent doglegs due to the offset of the joint components. The joints were still tight, though.

Wedges offer another variation for adding to the tenon's performance. Used on through-tenons, they can be driven in at the ends of the tenon, pushing hard against the joint elements, locking the two members together. If you can't do the through-tenons, then you may need to use a blinded wedge or fox wedges. Two names for the same joint, this one demands careful layout and execution. The strength comes in as the tenon is driven home, driving the wedge into the tenon, making it tight against the end walls of the mortise. Too long a wedge, and the joint will not drive home tight. Too short, and the mechanics cannot take place, and the joint receives no additional reinforcement. Of course, the joint cannot be taken apart if it tightens up and does not pull tight to the shoulders of the tenon. It must be right, and it must be just once. Samples will help get it right. Never a dull moment.

We will use double tenons on occasion. Driveway gates are often only about four to five feet tall, but will reach out to eight or nine feet, putting a lot of compression into the lower tenons, and lots of extension on the upper tenons. Racking. Double tenons will obviously double the glue surface and make for a very strong joint. Rarer to use them on a door, but when the job calls for it, it is good to have the experience and know where to land. Plus, there are few things as satisfying as driving home a double tenon joint. Perhaps twice the satisfaction of a regular joint.

Rather than try to use the same dimensions each time you set up the joint, just let the project determine the best dimensions. Length may be the same each time, with door stiles being the same width from door to door. Width will vary with the rail and parts to be joined. While the thickness will be determined by the tool and an effort to make the tenon 1/3 the thickness of the parts. The mortise and tenon joinery for a door is now second nature, given little thought unless substantially different in some way. Joinery for a door from here typically falls into two camps, coped and square edge.

Tooling

Tooling is the generic name for all the cutting tools in your shop: blades, knives, heads, and more. Sets of cutters, as opposed to one-piece insert-type knives, are often used to create the entire sticking and cope profiles for our doors. The sets offer the ability to take them apart and just use components of the set for what is needed. The versatility adds complexity, but that versatility comes in handy when you have to make a 2⅝-in.-thick door, or change stickings halfway up a door.

Over the years, we have made hundreds of different stickings. Some were to match historic profiles, some were by architect's design, some to match existing work, some just for whatever. The reason is no matter. What is important after the build is to label your set-up parts, and keep them for future reference. This will be a great real-world reference when the need arrives.

Looking at the sample shelf with most of these profiles, I ask myself how many times I need that particular thing again after doing one job. Did the second job (or more) ever come in? In most cases, no. But I feel the need to store the parts and info since the energy, effort, and head time is already in that solution. I can save some of that precious brain time by finding a solution on the shelf.

The photo shows the cutter set for our usual 9/16 in. ovolo sticking. The colored discs are fractional spacers, with each color designating a certain thickness of shim. The shims can be added or deleted depending upon the wood, the size and number of the joints, the glue, and how assembly is anticipated. White oak will require a slightly "lighter" fit since the oak will have no give. When joining pine, we tighten up the fit to let the pine crush a bit under the clamps. Either way, the goal is the same: to make a cope joint where it is impossible to see if it is a cope or if it is a perfect miter.

Appearances aside, that tight cope joint has a real-world function with water repellency. The tight cope will not allow water in, and it will protect the joint in the wood. Glue the copes, by all means, but it requires judicious application to avoid making more work.

The cope cutter set is of a larger diameter so it can make tenons up to three inches long. Both copes are loaded onto the spindle with the colored washers deemed appropriate, and the set is spun in the shaper. We use a custom-made air clamp sled for the end work on all our rails. It is safe, durable and not overwhelmed by some of the large or odd things we need to cope.

Tooling for the raised panels is a pair of three-wing cutters—called the "Innovator"—that are six inches in diameter, and have inserts made of corrugated HSS, three per head. One head runs clockwise; the other is upside down, and runs counterclockwise. They can both be mounted on the spindle and can raise panels on both sides simultaneously. We typically raise on the upper side only, since we do not have duplicate cutters of both "hands." We have bearings to mount under the cutter to prevent the panel from "overcutting" if it jams in the machine. We have hand ground our own profiles for the

panels in hip, ogee, step, and cove profiles. Having that capability alone enables us to match other makers' work if we need to.

The hollow chisels and Maka cutters make 99% of the mortises in the shop. The hollow chisel cutters get sent to a sharpening shop. Hollow chisels are, by their nature, abused. They are often overheated from a too-close fit to the drill, and/or a too-aggressive feed rate. The Maka cutters no longer have a source for sharpening. We can do a quick touch-up with the Maka, but it is nothing like coming from a shop that knows how to hold the tools for grinding. We can slow the feed rate so as to not overheat the oscillating chisels, but too slow and they burn anyway. Finding new chisels or getting these ground correctly will be a challenge since they have not been made for two years now.

Rabbets and grooves for the doors can all be made with the cutter sets, with cutters set in to remove sticking for glass panels, or with additional heads. We use carbide chipped groovers that can go from 1/8 in. to 1/2 in., and from 1/2 in. to 7/8 in. They can also work at rabbets, of course. In addition, there are two aluminum heads for pattern shaping and general rabbeting. These have matching diameter bearings, making them ideal for flush cutting to a template, and more. The better European tooling can regularly run at 10,000 rpm. While the smaller cutters and hand ground things don't go much above 4,500 rpm or so.

With my long experience in the shop, we also have the old split-collar beveled edge knives that were all there was until about 25 years ago. They come in several widths and thicknesses, with 5/16 in. and 3/8 in. being the more popular thickness. These knives (several hundred pairs) have to be organized in drawers so we can find what we need. With so many knives, we will just need to find them when the time comes. These were once considered inherently dangerous, but if one follows simple guidelines, the knives cannot come out. We are also fortunate to have a large collection of about 120 three-wing HSS cutters for a small shaper with a 1/2-in. spindle. This machine is miles away from the big SCMI shaper, but it does some things the big one cannot. The shaper—a 1940s Rockwell—has a spindle insert cartridge that allows the cutters to be mounted and fastened with a cap screw to the top of the spindle This then allows a tenon to pass over the entire cutter head. This enables longer tenons since we can cope with a router. We often can find everything we need to make a casing to match some arcane molding. While this stuff sometimes gets in the way, it is invaluable to have all these profiles ready to go.

It is not unusual during the early part of a project, to wander about the shop, looking for this or that profile, or a way to cut this, or another way to make that. Collecting shaper cutters, corrugated knives, a router bit or two, even some of the three-wing cutters. Now is the time to see what can be done, and how we will do it. This is problem-solving. This is what we do—problem-solve—we just do it in wood. Once we have a solution for all the needs of the project, we can walk it out as needed for a smooth, well-planned project. No surprises.

Hand Tools

Hand tools play an important part in making doors. As projects get larger and larger, it becomes apparent that we are better off taking the tool to the wood rather than the wood to the tool. So hand tools mean powered and non-powered to us. While a door can be built entirely with hand tools, it is rare to see this done for profit, as it is hard to get paid for all the hours it would take. Nevertheless, hand tools play a part, and deserve our attention.

When busting out lumber to rough length, we used to use an 8¼-in. circular saw. You know the drill—grab the saw from its place, find an open end of an extension cord, plug it in. Be sure you have power, then go cut the board. We bought a cordless DeWalt circular saw that will cut to 2½ in. thick. Just thick enough to go through the 10/4 mahogany we use a lot of. It has sped up and eased that entire process. No cord to fetch, plug in, get out of the way, flip over, turn around. Just cut the wood and move on. Very direct.

Combination squares are used to lay out the stiles and rails. Find or buy a good make, and check it for square before buying. Regular machine squares can also be used. And even framing squares can be used for a batch of stiles. Just know your squares, and how they behave in use.

Hand planes do have a place in the modern shop. We use block planes and rabbet planes more than nice Jack planes and joiner planes. One use is to ease the tenon ends for an easier fit into their mortises. A few passes, one side or both, at a slight angle from flat, will impose that angle on the tenon end. More than a few passes, and maybe we need to go back to the tenoner. Block planes find uses every day, every project. We keep a few sharp and ready to go. Once picked up early in the day, they remain on the bench all day.

We are fortunate to have been gifted two full sets of dead-blow mallets. From a small four-ounce item to a nine-pounder. Fortunately, the nine-pounder gathers dust, but all the others are used in assembly to seat or coerce parts into place. Wood scraps will be employed as needed to protect the wood from dents from the mallets. We save the ripping (about 3/8 in. by 2½ in.) from exterior jambs to use as stile pads to prevent damage to the stiles as we clamp up.

We isolate the door parts on the bench by elevating them on bumpers, as we call them. These bumpers are tall enough to allow clamps to pass underneath without pushing the door parts around. Every door is glued up on these bumpers, so that the bumpers, not the doors, get a lot of the glue and anything else on them. They need to be parallel dimensions so they extend the flat reference we have established with our benches. When there is more than one door to assemble, another set of bumpers is employed to raise it on the door just assembled, again preserving flat. We also keep a few stacks of blocks at 1¾ in. by 3 in. by 4 in. These can used in place of bumpers once the door is assembled, again using them to preserve flat and never allow twist.

Pipe clamps are the usual clamp put to work pulling up doors. We like them since they are relatively light, but can really pull up a door. Placed above and below the joints to avoid inducing a twist in the rail/stile joint, the two clamps can pull together and bring the joint into alignment. We also have bar clamps. These are put to use on larger, more reluctant assemblies, or brought in when the pipe clamps are not pulling things together.

Hand-screws—those odd clamps that are seen more often decorating walls—have a place in the modern woodshop. Learn how to open and close at will, and they suddenly become more useful. We use them for many things, but thickness lamination is one of them. Learning to lever them into the lamination is an acquired skill, but once you learn, you realize how much clamping pressure these things can supply.

Glues

Modern glues are varied enough to cause confusion for the woodworker needing a reliable glue. There are PVAs, PURs, PRs, aliphatics, epoxies, and resorcinol. There are rigid glue lines and thick glue lines and thin glue lines. Fast-setting, slow-setting, mixed or one component, thick, thin, visible, or not. All wood glues are strong enough to

break on the wood rather than the glue line if applied properly. So strength is not a factor in our decision.

Interior doors can be glued with the preferred aliphatic resin or "yellow glue." The demands on the glue lines are just much lower with a nice, dry, stable environment. PUR or polyurethane glue is also a good choice, especially if you need more open time. PVA, or white glue, can be used, but it has largely been replaced by the aliphatics since they have better sanding characteristics. Learn to work with more than one glue so when you do need to use it, it is not a foreign substance to start out with.

Gap-filling is a property of some glues. Enough of any glue will fill gaps, but most are not structural in the gap. That is, they do not bond where there is a gap, though they may well be in contact. Urethane glue, with its foaming characteristics leads many into thinking it is a structural, gap-filling glue, but it is not. It merely expands as it cures. Resorcinol has light gap-filling properties. Epoxy has the best gap-filling properties we see amongst the common wood glues. In fact, epoxy joints should a bit loose since epoxy likes a thicker glue line.

Use the disposable 1/2-in. to 1/4-in. acid brushes so you can get in and coat the sides of a mortise with glue. Paint it as you need to—up onto the sticking for the copes to bond, throughout the mortise and the haunch. If you have several rails and parts, you might consider a slow-set aliphatic or urethane glue. Either will give you extended open time so you do not have to risk having parts bond where they should not, or glue drying before it is mated to its other part. Knowing a bit about more than one glue will pay off when it comes time to use another type of glue.

Exterior doors require a different approach to the glue, since the exposure—or amount of exposure—matters. We have learned to avoid Titebond III in all cases. The problems we had with TBIII are detailed elsewhere in this book. Some exterior doors are sheltered by a roof or, even better, a northern exposure, and this means the door can be glued with TBII or even TBI. Often, we may not know the details of exposure, so we have to assume the worst. That means a glue that is rated as Type I Exterior. We also like a rigid glue line so things cannot creep over time. Some doors just have a few mortise and tenon joints, so epoxy or TBII might be the choice. Some doors may have just the basic mortise and tenon joints, while another door may also have bolection molding to attach, and raised panels to laminate, and a more complex glue profile.

As of today, subject to change, we use West System epoxy for exterior door joints—glue for width, laminate for thickness, panel laminations, including face and cross-band veneers and core-to-solid joints. Solid wood panels will use epoxy or TBII.

Forty years ago, choking through some dusty resorcinol mixes, we used to dream of a waterproof, one-part glue in a squeeze bottle. Since then, two glues—TBIII and urethane—have arrived with that promise, and both have found some use, but they are by no means universal.

Chapter Five: A Simple Door, a Difficult Material, a Coped Door, Sashed Doors, Bolection Doors

A Simple Door

A simple door will be square-edged with no profile, and be just five parts: two stiles, two rails, and a wood panel or glass. These are sometimes called "open light" or just "open" doors. Regional variations on woodworking terminology are rampant, so you may have heard other terms for this basic door.

First, the cut list. While it may seem silly with such an uncomplicated list, enjoy it while you can. Cut lists can easily run several pages, and be very complex. And do not push off the need for a cut list for something so simple. Working the same/similar visual format every time helps create expectations and establishes good habits. If there is a question at any time, the cut list will have an answer. Mark the board where the rough stile lengths will be, and make those cuts. Then do the rough rip, adding enough to plane or joint off for straight. Same for the two rails. Pick up the rippings off the floor to make the stops to hold the panel in place.

Then, face and edge at the joiner, using all those newfound skills. You might warm up with the rails, then face and edge your stiles once you are at operating altitude. Then, on to the planer to go to the final thickness. Run the width at this time also, for the stiles at least. Rip the width for the rails. Top and bottom rails are often left an inch wider than the final size in the completed door. This is so we have plenty of wood to trim fully and make the dimension we need.

Remember to add the face marks once all the sizing is complete.

Next is layout. This step should not be passed over, since it still does several things. We will place the rails into the stiles accurately only with a good layout.

The panel in our simple door will be used to exemplify the set of characteristics that makes wood the unusual

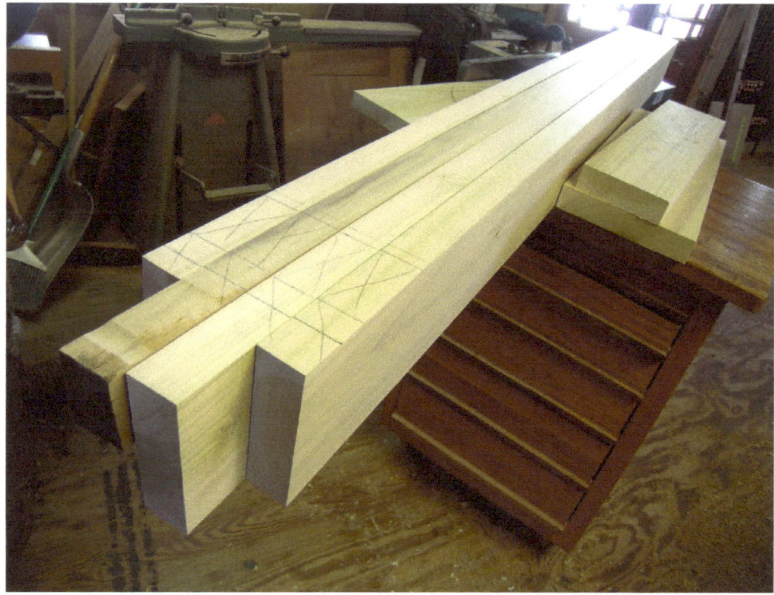

material it is. Much mystery and lore surround the hygroscopic properties of wood. Coupled with the fact that wood is not isotropic—the same properties in all structural dimensions—things appear to get difficult. A board will change in thickness and width with changes in moisture content, but not in length. Wood is anisotropic. This can really complicate things, but we know how to deal with it.

The panel is to be sized according to its construction type. Assuming it is plywood, MDF, particle board, or some other man-made, stable board, it will not move in normal conditions with changes in relative humidity. The panel should be undersized to the opening. How much undersized is the perennial quandary. "Opening" being the bottom of the stile or rail plows once the door is assembled/dry fit. If in doubt, undersize it a bit more. To be sure, dry fit the five parts together so there will be no hitches once the glue is spread.

A Difficult Material Explained

Wood is warm and smooth to the touch. It is the only thing that looks like wood, yet it is impossible to duplicate or improve upon. However, it has this characteristic that has caused woodworkers for centuries to pause and consider: Wood moves. It expands and contracts according to its environment. It never changes length with changes in humidity. It just does not move that way. It might be helpful to think of wood as being a bundle of drink straws. The bundle does not change size, but the straws can hold a lot or a little of water. Regardless, the shape stays the same, swelling slightly as the straws fill up with water. The length will not change.

If the panel is solid wood, then it is crucial that expansion room be added to (in reality, subtracted from) the clearance size of the panel opening and the actual panel size. But only for the dimension that is across the grain. Since wood expands and contracts across its width and thickness (anisotropic—wood does not functionally change in length with changes in moisture content or relative humidity) the clearance needed for the panel width should be calculated. Thickness can be ignored since a moisture change will not occur in 1/2-in. material.

If the panel is too wide, and expands in service, it will stress the joints, and could even cause a failure. If the panel is too narrow, it could slide over to one side or the other and show a gap. It is best to do the calculations until one has enough experience to feel confident sizing as you go. The panel length is not so critical, and can be the same size as the opening length, allowing the panel ends to bottom out in the rails. It is also good practice to center the panel in the opening.

For centuries, woodworkers have had arcane, secretive and/or non-existent rules for determining the movement of panels, and therefore how much room to allow. Today, we have a good accumulation of wood research, so gadgets like the Shrinkulator are readily available for calculating clearance (woodbin.com/calcs/shrinkulator).

In my experience, when wood arrives from the vendor, it is at its driest. It will only acquire moisture from this point on—in manufacture, transport to another environment, and in service. This is true with interior projects and exterior projects. So, if faced with panel sizing, I can rule out panel shrinkage from the point of building since it is likely to occur only in unusual conditions of high humidity.

Most of us who have been faced with jobs where completion is more important than how it is completed may have heard ourselves say, "I'll just glue it real well so it can never come apart." But wood expands, and nothing much can stop it. That ability to expand was exploited by so-called primitive people to quarry stone using bits of wood. A series of holes would be drilled along a line parallel to the rock face. Dried wood wedges would be driven in to the holes, and then soaked with water. Each wedge absorbs some water, then expands—or tries to expand, putting pressure on the sides of the hole. A line of wedges exerts this pressure all along the line and eventually, a crack develops. More wedges, more water, and in time, a pyramid is built.

If we carry this thought experiment a bit further, we can better understand this property of hygroscopic movement in wood. Imagine we have a tabletop at 30-in. wide and 6% moisture content (MC). The relative humidity (RH) is increased, and the tabletop absorbs moisture and expands, changing the 30-in. dimension to 30⅜ in. As long as the construction of the table allows for this, all is well. If the RH is decreased, then the table dimension may well return to 30 in. This is seen in seasonal moisture changes in many wood products.

Next, imagine two concrete pillars at 30 in. apart. Our table top is just able to fit snugly into the gap. Then the RH is increased, and the top is absorbing the moisture. But the pillars restrict the movement, so the top cannot expand. This is the equivalent to "gluing it real good." But what is unseen at the moment is that all those wood fibers are crushing just a little bit in order to accommodate the increase in moisture. This is not seen under normal conditions. Now, if we were to reduce the RH and take the top back to its original 6% MC, then it will be less than 30 in. wide. Why? The crushed fibers have lost their ability to swell back up, and the top remains in this "compressed" state, even through subsequent seasonal changes.

One more step: Think of the 30-in., 6% MC top being placed between the two pillars again, but this time we will glue the edges of the top to the pillars. This would be an attempt to prevent the top from the "shrinking," as seen in the previous example. Equal to the phrase "We're gonna glue 'er real well." So, now we add to the RH, and the top "tries" to expand, but cannot due to the pillars. We let this sit for a while, and then the RH is reduced. We would expect the top to be less than 30 in. based upon the recent example, but the glue prevents this. At some point, we hear a shot, or maybe just a tearing, a rending, as the top opens up a split or two. It could split into two pieces if the MC was extreme enough. The top could not "shrink" as its nature dictates, since it was fastened to the pillars. But the fibers expanded and crushed when the MC was increased, and then decreased in size as the top lost MC. With the top restrained by the pillars, that shrinkage has to go somewhere, and splitting the top is where it appears.

Careful examination of cross-grain wood failures will almost always be explained by the foregoing. But it is important to understand why and how wood moves since doors have panels, and not all of them are going to be plywood.

A Simple Door, Part II

Now, a quandary of sorts. The cut list is filled, and our parts are S4S'd. Laid out and ready for joinery—tenons and mortise, but which goes first? Which one is the chicken, and which is the egg? This choice is yours, but it will be driven by which of the two elements is adjustable. The type of equipment and methods you have to make mortises and tenons will dictate the better sequence. That is, if your hollow chisel mortiser is to be used, that cutter is one size, and one size only—no adjustment. So, it will be the chicken. Or egg. Either way, it is made first, and then the tenon will be made to fit. We use either a chisel (Maka mortiser) that is the exact dimension, or a hollow chisel that is also the exact size, so mortises first and tenons to follow is the more common sequence in our shop. A 1/4-in. hollow chisel will often make a fuzzy mortise, especially in softwoods. The width will be about 0.285 in. with soft woods. A full 1/32 in. larger than the tool. This would call for the mortise first, then the tenon should be fit to it.

Make a trial mortise in a piece of scrap wood the same thickness as your planed door parts. Then make a trial tenon. With our equipment, we center the mortise first, then fit the tenon, then center it all to match.

The tenon should be centered in the thickness on the ends of the rail, so the mortise will also center in the thickness of the door. The fit of tenon to the mortise is critical. Make it so the joint will just stay together on its own. We will assemble the door dry, with no glue, to check assembly methods and see if there is anything unusual, and we can pull a tape to check width and length to be sure we have it right before we spread glue. Dry fitting more than one door requires taking the dry assembled door off the bench and standing it up next to the bench. The loose, dry joints should be snug enough to allow the door to be stood on its end, and remain together. A falling top rail tapping you on the head is not the way to learn the tenons are too loose.

The depth of the mortise, beyond the basics of the joint, is determined by the types of equipment making the joints—and their limitations, if any. A normal 1¾-in. by 36-in. by 80-in. door will do nicely with 1½-in.-long tenons (plus some dimension for sticking—if any). Shorter tenons will give less glue surface, and longer will give us more glue surface, but the additional length is not needed, though it is reassuring. If you like the security of long tenons, then mortise away, and enjoy.

The mortise should be about 1/8 in. deeper, to allow the tenon shoulders to come up tightly. The end grain on the end of the rail does not count in our glue surface calculations, and we do not count on it for the strength it might lend to the door. This void also can take the excess glue should there be any. A large excess can go out the tenon end and follow the haunch to escape out of the joint. That is your cue to review how much glue needs to go into the joint.

We will next make the plows in the four frame parts. This will require a 1/2-in.-wide plow, about 1/2 in. deep. Or, if we have a panel of a certain sort, the plow can be sized for that specific thickness. The stiles and rails all get plowed the same way, the same time on the same settings. And with the face marks all up, or all down. While we have made the effort to center the tenons and mortises, and may well have it dead-on, we still run faces up, or faces down through the job because it helps us keep things straight when things get more complex.

Once the two joint components are made, then we need to haunch the tenons. The haunch is a shortened form of the tenon that performs some tenon function, but at a shorter length so as not to leave the mortise open to the elements. The mortise should enclose the tenon.

As a rule, there are not too many rules on haunch sizing. None, in fact. So, enjoy a little creative freedom here, and make the haunches reasonable. They should protrude a bit more than the the plow is deep, and they need to be no less than 1/2 in. from the top end of the door, and about 1 in. to 1¼ in. from the bottom of the door. Haunches should be marked at layout. The mortiser can be reset for the second depth, or the plow can be made deeper here by hand feeding in the stile at each end at the deeper setting, just far enough to level out the haunch so it is parallel with the main plow floor.

Assembly will begin with the bench being cleared, and two suitable door bumpers placed on the bench. The bench is sturdy and leveled so we can "start flat, stay flat." The bumpers will elevate the door off the bench and allow for pipe or bar clamps to pass under the door. With our beam clamps, we use 2½-in.-tall bumpers. One clamp on each side of the door, at each rail, should be adequate to pull the joints up. A mallet (dead-blow type preferred) is useful to coax the rails into place on the layout marks. Use a block of scrap to prevent marring any surfaces seen in the finished work. A tape measure will measure for equal distance inside the two rails, or a pair of sticks can make this measurement.

We like the 1/2-in. disposable brushes with black bristles in aluminum tubes. They can be reused, but if overused, can lose their bristles all in a clump. We can use a yellow glue for this door. The regular glues will work fine if the door is to be used indoors. If in the weather, or at risk, then we would step it up a bit to the second yellow glue that is "water-resistant." Squeeze glue right into each mortise and then go back around and brush it up along all four sides of the mortises, and down the haunches, too. If you get waylaid somehow, then go back and brush that glue up the side walls again, making sure there is good coating. Then roll up onto the edge of the stiles, and brush a good coating along both sides of the mortise and the shoulders. The goal is to have all surfaces of the joints in contact and glued. Hold back a bit from your layout line so there is not a lot of—or any—squeeze-out at that point. As the joint comes together, that glue will spread out under the joint. The goal is to stop just short of it squeezing out of the joint, necessitating cleanup of some sort.

Glue at the junction of stile/rail/panel can be a serious problem if it restrains the panels from normal seasonal movement. Solid wood needs to be allowed to move with changes in relative humidity. Holding it back is critical to avoid a glued panel. If the panel is a man-made board, then it can actually be glued into the frame. Glued all along the plows if you like. But solid wood must be allowed to move, with expansion being the most likely direction.

Fit the top rail into a stile sitting on edge on the bumpers. Align the face marks. Guide the rail into the mortise so it lands on or near the layout mark. Use the mallet if you need to, and then drive the rail in almost tight. Repeat with the bottom rail. Pick up the other stile and place it on top of the rails standing upright, and fit first one, then the other tenon into their respective places. This is where a second set of hands would be helpful.

Lay the assembly onto the bumpers, check your rail placements, and tune as needed. Pick up a couple of rippings that are about the same width as the thickness of the door, and about the same length. They need only be 1/4 in. or thicker. We collect the rippings from the exterior jambs when the rabbet is cut. This yields a scrap about 3/8 in. by 1¾ in. or 2¼ in. Perfect to protect our nice door stiles. These side pads will protect the stiles from mallet, bar, or pipe clamp damage. It is handy to build a collection of pads as well as bumpers. Keep both assembly aids clean and free from wet glue so as to not inadvertently glue something where you do not want it.

As the clamps are placed and tightened, inspect the layout lines to see how they are staying in alignment. Wide rails will be self-squaring, but narrow rails may get out of square, so it is wise to check diagonals if there is a chance of the assembly being out of square.

The door must be clamped immediately after assembly to allow the glue to set and hold everything in its designated place. Clamps are opposed to each other on either face of the door, in order to even the pressure on the joints. Use a combination square or straightedge to ensure the stile and the rail are level, coplanar. If they are not quite right, adjust a clamp to position the parts correctly.

In the heat of assembly, it is easy to forget something, or get it wrong somehow, so practice it first with a dry run, then go slowly enough to assemble without incident. If you wish, you can clean off excess glue where you can, but the underside will be difficult to clean.

Once the glue is set, clamps are carefully removed and side pads put away. Scrape the joints to clean off much of the glue, and use the scraper to level any joints that may be out of level. My favorite tool is the Red Devil Paint Scraper #3050, 2½ in. wide. Sharpen the blades with a bastard file, and the tool is fast and clean-cutting, good for removing glue and leveling joints if needed.

The door can be left upon the bumpers and prepared for sizing. I cut length first, marking the top cut line and the bottom, and then measuring again. This is no time to save a few seconds by omitting a second measurement. We use a Festool track saw and guide to size the doors. Use a good square to do the layouts and cut the line. Check the shoulders of the rail/stile joints to clean up any squeeze-out. If you have done your work well, there should be no need for wood fillers, glue, and sawdust, or any other thing to fill gaps, etc. The door should be flat, sitting on the bumpers solidly, with no rocking. If you have a panel saw, then take the door to that tool and make your cuts. Panel saws excel at making simple square cuts of a good size.

Glue suitable for doors is almost as varied as the designs of doors. The aliphatic resin or yellow glue is a favorite for interior work. An exterior rated aliphatic glue has been found to be problematic for exterior work since it loses strength when heated. It also has a high "creep" factor and will allow bent laminations to straighten themselves out, among other things. The choice of glue is not related to strength, as any real wood glue will, when used correctly, bond wood as strong as the wood itself.

In our simple door example above, yellow glue would be ideal if it is not in the weather. The quick dry time is an advantage, and it dries sufficiently hard so as to prevent creep of any sort. Near panic is the only way to describe having to whale on a rail that has set up too soon and no longer will move without a grand show of force. A more complex design would prevent the use of the simple yellow glue since the open time is shorter than our anticipated assembly time. Urethane glue will have a longer open and close time, but is somewhat messier to work with. Epoxy is the slowest cure, and is the only gap-filling structural glue we commonly use. It has replaced all our use of the "three" product that gave us thousands of dollars of reworks and replacements. We warrant all our work, so we have to be sure of the glue and other things that could come back to haunt us if we are not diligent.

A Coped Door

A coped door build will be much the same as the simple door above. The fact of the copes makes for a more complex measurement across the width, since the copes need to be "gained" in the rail dimensions and deducted from the stiles.

We have an ovolo profile that is used most often with 1¾-in. doors. This profile is a quarter circle with a step above and below. It is 9/16 in. wide, enough to cover insulated glass spacers when used with insulated glass. It provides complex measurement across the width, since the copes need to be "gained" in the rail dimensions and deduction is appropriate in everything from Colonial to English, to transitional designs. We must account for the sticking (male component) in our stiles, first at drawing, then at cut listing, and also allow for the copes (female component).

If we need 5 in. on the stiles and top rail, and 10 in. on the bottom, then our cut list will look like this:

- 2—1¾ by 5⅝ by 80 in.+ stiles
- 2—1¾ by 5 9/16 in.+ by 29 in.— 26 in. Circle # with 1½-in. T.E.E. Top Rail
- 1—1¾ in. by 11 in. by 29 in.— 26 in. Circle # with 1½ in. T.E.E. Bottom Rail

Each stile "gains" 5/8 in. to 9/16 in. for the sticking, and 1/16 in. for the full-width cut to clean up plows and all the sticking. The tooling in this case will remove 1/16 in of the stile width in a full thickness cut, with fences offset for 1/16 in.; 5⅝ in. less 1/16 in. full thickness cut, less 9/16 in. sticking, 5 in. stile functional width.

The rails will gain 5/8 in. on each edge. Top and bottom rails will only gain 5/8 in. once. Mid-rails and internal stiles will have two gains of 5/8 in. Muntins also have two gains, plus the 1/16 in., so a 1/4-in. muntin will gain 5/8 in. twice—to total 1½ in. on the cut list.

A good shaper can have its fences offset by 1/16 in. in this example, and operate like a joiner—taking off a full depth cut at the beginning of the part, through to the end. Our set-up parts will help us see if it is set up

correctly so the part does not climb out of the cut, or is set too deep and will not make it to the out-feed fence.

The tooling to make our 9/16-in. cope and stick joint was designed for flexibility. The sticking set includes several plow cutters to stack so it can be used on doors from 1½ in. to 2½ in. thick, or more. The top and bottom cutters are separate cutters. One can be removed, and a glass rabbet cutter inserted. A set of spacers that runs from 0.003 in. to 0.030 in. is color-coded so we know what thickness each shim is by color, without measuring.

The cope cutters are similar in that they were designed to do anything the sticking cutters can do. They are brazed tooling like the sticking sets, but of a larger diameter so we can poke a tenon between them on the shaper. Larger, more intimidating diameter. The shaper spindle is 1¼ in., and the spacers that separate the two coping discs are 1/2 in. wide each, and the tenons are 1½ in. long, plus the 9/16-in. cope. The discs must be a minimum of 7⅜ in. Ours are close to 8 in. so we can do 11/16-in.-wide sticking. The discs are spun at about 5,000 rpm, so the tips are moving at about 120 mph. They nearly vanish as they reach speed. This requires a stout shaper, with a 1¼-in. or wider spindle, several hundred pounds of cast iron to soak up vibration, and 6 hp or more. Sandbags in the shaper cabinet can add vibration-damping mass.

When we need a 2¼ in. with 9/16 ovolo sticking, we go to our set-up sample and find our notes on the sample. These notes instruct us as to the best set-up, and what spacers are used where. These notes can be updated with footnotes for pine versus oak, and how they differ in set-ups. We do not encourage too much experimentation, as this is fundamental to what we do, and should be fast, easy, and reliable.

The overriding design factor I had when I designed our tooling sets was versatility. This was achieved by breaking up the cut into four, five, or six cutters. If my objective was ease of set-up and speed to get the first part out, I would have insert cutters made onto one cutter body for 1¾-in., and another for 2¼-in. doors. The thin carbide knives would be ground to pattern, and were as replaceable as a whole pattern. Changing to sharp, new carbide cutters would only take a few minutes, and set-up would remain unchanged, but much faster.

The insert cutter tooling now has great appeal to me, and I will move in that direction as time and need direct.

For someone starting out, I think I would recommend the versatile brazed sets for both cope and stick, since they can be used for so much. If you have a firm grasp of the type of work you will be doing, then the insert sets may be the best place to start.

We also added bearings to our order, so we can run curved parts with a flush bearing for both sticking and even copes. The sticking is used to profile curved rails reliably and repeatedly. The rail can be rough-sawn on the bandsaw at about 1 1/16 in. to 1/8 in. over size, then run on the shaper, and a clean, complete, full profile cut made. If the parts drift off the bearing, then it can be run again.

Tooling companies are not all the same, so be sure the one you select has experience with cope and stick cutters, and will test them on a piece of wood, guarantee the fit, and sharpen/supply inserts as needed in the future. Custom tooling sets can easily cost a couple of thousand dollars per set, so choose carefully. Draw out the cuts full size, both cope and stick, and describe the rotation, spindle size, and horsepower of your machine. Visualize how the different cuts would be made, and ensure that the bread and butter cuts—the everyday plain Janes that really pay the bills—will be fast and accurate.

In order to round out the description of tooling sets, it should be mentioned that, back when Delta was Rockwell, they designed a light shaper for cabinet shop or home use. This had a standard 1/2-in. spindle, and several other diameters were available since the machine made it easy to change them out. A stub cutter received three-wing, high-speed steel profiles and retained them with a recessed screw that tapped into the end of the spindle. This allowed for copes of an infinite width. That will never be needed, but if you need to make through-tenons for a coped door, this set-up allows the tenon itself to pass over the cutter, while the tenon shoulder is coped.

We had our cope patterns made into small cope cutters for our 1952-model Rockwell shaper and had the sticking made into matching router bits. We can also run these bits with bearings below so we can do curve copes and repeated work. There are many other sets of cope and stick that can be run on this machine, one cope, one profile at a time. The panel-raising heads and loose HSS corrugated knives can be employed to hand grind a profile that can then be used for matching old work, or for something new. These options are rarely used, but we have them when we need them, and I feel it's important to offer a wide selection, just to be able to call our shop custom.

Let me explain that term "hand grind." The more expensive pattern grinders can, with extreme care, grind four knives in a molder head so each knife has the same exact projection, all within a few microns—0.0003 in. This is desirable since all knives will be in the cut, doing their part equally, prolonging tool sharpness and giving a quality surface that does not show mill marks from the cutters. This good knife cuts per inch (KCPI) is what makes for a molded part that needs little or no sanding.

However, consistently achieving knife projection values as above is nearly impossible—and certainly impractical—for us in our shop. The planer has four knives, carefully set, but only one knife actually makes the finish cut, as its projection is the greatest. Planers with grinders that can joint the knives on the fly can get better feed speeds, but the resulting flat on the knife tip will increase the noise level noticeably and sap cutting power. You can use a hand lens to see and count the KCPI from various machines in your shop. Carbon paper rubbed on an area will highlight the knife cuts for easier counting.

We have a feeder that runs about 20 feet per minute. A three-knife brazed cutter is placed on the spindle and spun up to 5,000 rpm. That works out to 15,000 knife cuts in one minute. But since only one knife is creating the surface we see, then it is more like 5,000 knife cuts per inch/minute. At a feed rate of 20 feet per minute, then, we will see 5,000 knife cuts in a 20-foot-long piece of wood. Or 250 knife cuts in a foot, or 20+ KCPI. This is more academic than practical, but it is helpful to know what the KCPI should be within certain norms. For example, 18–20 KCPI is considered a good finish in moldings, but 12 KCPI is low grade millwork. Generally speaking, it is desirable to not see any KCPI, and to have smooth, chatter-free, mill-mark-free material, foot after foot.

As for our coped door, the parts are all ready, mortised, tenoned, coped, and stuck. A coped door goes together at dry fit, and assembly is the same as a square-edged door. Before gluing, check your copes with a full-up dry fit so you can be certain the door will pull up. Once the glue goes on, there is no turning back. It is crucial to glue the copes in these doors. Brush a bit of glue into the copes, staying away from the exposed ends, where squeeze-out is especially hard to clean up. Hit your layout marks and check for square as the door is clamped.

Clean up the excess glue with a damp rag or similar. There should be glue squeeze-out all along the rail/stile shoulders.

It is, as explained earlier, critical to glue the copes to prevent water absorption. Some years ago, a national maker of exterior doors placed 8, 10, or 14 of their doors in every Chipotle restaurant. These were open doors, coped, with one large panel of insulated glass, two stiles, and two rails. These were used in place of the ubiquitous black aluminum and glass prevalent in commercial architecture, a nice touch. The warm tone of the natural fir meshed well with the restaurant's principles of freshness and quality. After a month or two, I noticed that the stile/rail joint at the lower edge of the bottom rail was cracking the finish on the exterior. A slight bit of discoloration meant the bacteria were in there, waiting for more water, feasting away. I could easily see the open joint had never been glued. This maker of doors has a fine reputation, but the doors are doweled, and glue is applied only at the two or three or four dowels in each joint. The copes do not get glued due to the difficulty in removing excess glue. That failure helps production but is the first step that a wood door takes on its way to the dump. In time these cracks opened up the whole width of the bottom rail on every door panel. As the discoloration grew worse, I was curious as to how Chipotle would handle this.

Then, one day, I saw the result: not a new door with presumably glued copes, but that black aluminum storefront door that goes in quickly and easily. And now there is a corporate buyer of premium wood products that had a major failure and expense simply because no one likes to remove the glue squeeze-out from a coped door joint.

When I toured a large maker of doors some years ago, I could see all facets of a very good door plant. They had been making doors there for over a hundred years, so they knew what they were doing. As we watched assembly of the doors on hydraulic clamp tables, we could see the dowels dipped in glue and inserted, then the rails and stiles all pressed together by the hydraulic rails on the table. No glue on the end grain of the rails, and no glue in the copes at all. I asked my guide why.

She explained that when the copes are glued, there is a good chance of glue squeezing out onto the face of the two parts. This is a urea resin glue that is cured in five minutes in a subsequent step immediately after assembly. Once cured, the glue is hard to remove. These doors were ponderosa pine, and when chisels and knives were employed to remove the glue, it often lifted the grain, and it could pull out with the tools, leaving small voids that had to be filled. This was considered a larger problem than unglued copes in their view. The puzzle of how to clean off excess glue had not been solved in this venerable old shop. It was considered too difficult to train people to use various tools properly to remove the excess, so the problem was sidestepped by not gluing the copes. This is of course rendering ineffective the gain of additional glue surface desired with coped rails. One large gain is traded off for a slight gain, in my opinion. As I look at these doors, even today, I see the copes are unglued and at risk.

A paint scraper was mentioned a few pages back, with the simple door. I will mention it again, as it is the tool that, with a bit of practice, will scrape joints level and mostly glue-free, without damaging your door. First, learn to sharpen it at the bench with a good bastard file. File all the way across from end to end, with even pressure. This will help keep the blade straight. Once you can feel a good "cut" all along the edge, flip over the tool and do the other edge. Find the right angle on your wood by making a trial pass or two. The blade needs to cut the joint not only at the right attack angle—angle from door surface to tool—but also from above. The blade should pass over the area at a 45° angle so as to skew under the glue, and cut it loose, rather than snagging it and pulling it from the door. A bit of practice and care will pay off with some quick and efficient cleanup. Epoxy is the one glue that does not respond to this technique except in very small bits. Usually, it is best to load the sander with an old disc and grind off the epoxy.

The next point of entry for water in a wood door is in the same area on the actual bottom surface of the door. The stiles expose their end grain here, and the haunched cope joint is also there. The end-grain structure of wood once transported thousands of gallons of water to the canopy of the living tree, so while "dead," they can still wick up water by structure alone. The painter is supposed to paint the door bottom to seal up that point of entry, but they almost never do this. Acorn has adopted the process of painting all exterior sashes on the bottom of every door. This alleviates trying to get the painter to do his job, and presents a superior sealed surface that will last the life of the door.

In service, any door is likely to get water on the face, and it is likely to run down to the bottom, and wick around in that tight space between wood, threshold, and weatherstrip. In most of the U.S., freeze/thaw cycles will exacerbate the opening of the joint. We have felt for 30 years that it is imperative for any door-maker to do all in their power to produce a door that addresses any problems seen in the field. I am unaware if anyone has actually addressed this problem. Things like that just may be why the wood door industry is doing poorly. Innovation is gone from the industry.

Painting the door bottoms with epoxy is a big first step. Glued copes is another big step. As makers of a premium product, we have to do something to make our product stand out, something to make them unique. Unique in a world where cheaper seems to be the goal of every maker. So Acorn doors can stand out as a result. Subtle and simple things can make a big difference in the life of a door.

Sashed Doors

Sashed doors are simply doors with sashes in them. These sashes can be utilitarian, as for a window, with a replaceable sash for longevity, or decorative, adding an attractive layer of depth and molding. Sashed doors may hold panels in place of glass. Indeed, many sashed doors will have both panels and sash in combinations.

Building such doors is pretty straightforward, with a good cross section to guide the way. Often the "sash" can be any of several iterations of raised panels, or right angle ovolo frames, raised like a panel or rabbeted to fit into the larger perimeter frame at the door stiles and rails. Or, left square and made to fit into a 1¾-in.-wide plow in a 2¼-in. squared-edged door frame. This can give us a nice, wide wood area that will add to the door and make it unique, but without being too obvious.

Since this just might be the door to the king's counting house, this is where we stack the moldings and the depth and the sash in order to convey importance. A latch of equal import will impart the security that would go with such importance.

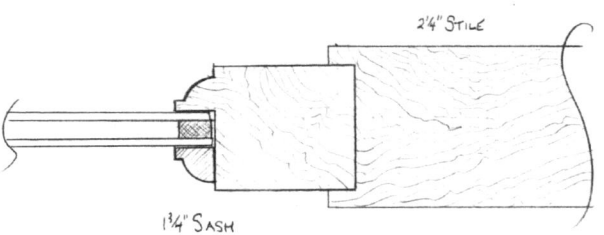

The simplest form of a sashed door is shown in section in the drawing. While not very dramatic, it is basic and is used in combination with other techniques to make for a grand door. Typically, one technique is adding bolection moldings.

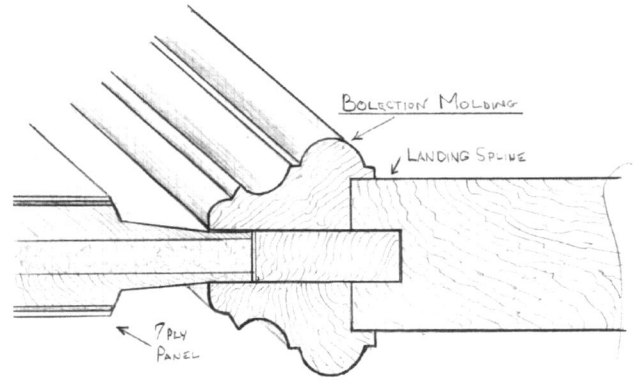

The lands that run under the bolection molding are important, though never seen, since they must be parallel to the faces of the door. These lands should be snug to the groove in the rails/stiles, so that glue can be run into the groove, and the land tapped into place and checked for parallel. Too tight, and they can twist as they are driven in. Too loose, and they are still able to twist. Get it right, and it will fit perfectly. Tap into place, and it seats into the bottom of the groove and will be true to the rest of the build. We have several types of bolection moldings, from 1 in. wide on up to 5 in. or so, with the patterns all somewhat similar so they can be combined and used as needed to help create the effect desired. It is important to machine this plow for the lands in one pass for consistent width and depth. If the width changes even a bit, it will throw the lands out of whack and create other problems. The purpose of the lands is to keep the molding in place, and to reinforce the miters with all the glue surface the lands provide.

The sash, which is secondary to the stiles and rails, can be made like it is another door. Mortise and tenon corners, with cope-and-stick-profiled inside edge, and the outside perimeter rabbeted or made into a panel raise to fit into the primary sash. The center can be glass or a solid panel. Or even another sash.

If divided light glass is required, then the inside edge of the secondary sash will need to be the 9/16-in. ovolo so it can establish the muntins and glass divisions.

Within the sashed format, there is a lot of freedom to add curves, mitered assemblies, emblems, and carvings. While the doors do not require much more than 10%–20% more material to be built, they can easily take two to three times as much labor to build. Doors in this class will be the ones that help establish a shop as knowledgeable and able to do about anything.

The challenge, when a customer is asking for such a thing, is coming up with a design that will satisfy the customer's desire for something truly unique. Google is your friend. You can search for old and grand doors and start to pick up elements of things you see and can incorporate into your project. Keep your project original, and don't lift enough to make for copyright problems, and try to stay fresh.

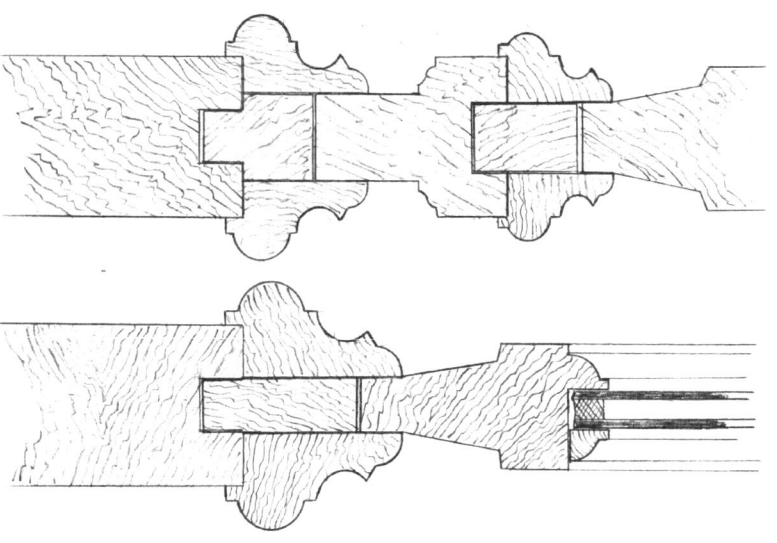

Bolection Doors

Bolection doors are square-edge doors that use an applied, or bolection, molding on both sides to retain the panel or glass. Most often a bolection molding protrudes above the face of the door. This dramatically adds to the depth of the door, with larger, heavier moldings used more often than thin or light molds.

We probably have 18–20 bolection moldings we can pull from for a project. Some have more depth, some have more width, and a few have both. Most of our work uses one of three moldings related to each other. The original profile was copied from a 17th-century cabinet made in Lyon, France, hence the moniker "Lyon mold" with the American pronunciation "lion mold." Each of these three moldings has a range it can cover when bridging from the face of the door to the face of a glass panel, a sash, or a panel raise.

Bolection moldings have traditionally been nailed onto the face or edge of the doors to retain whatever kind of panel was needed. Over time, the nails would loosen due to wood movement and the fact that the miters were opening. Once there is a gap on an exterior door, water will find its way in. It may freeze, pushing against two elements as the ice expands. Or it may just develop an area of moist wood that invites the various bacteria and fungi that enjoy such places. The exterior miters will fail first, then the interior miters will follow in time. For the upscale look the moldings give, the doors just are not long-lived if there is any exposure at all.

As I run across old doors with bolection moldings, I like to examine them to see what can be learned. I now feel that small shops probably produced these exterior doors, and did so without the cope and stick tooling nor ability to make coped doors. A square-edged door is much simpler than a coped door, but boring. Bolection molds can be nailed on, and the door is elevated. I think this was probably the way many shops produced upscale doors in the late 19th and early 20th centuries.

So we have a door type we need to be able to reproduce, but without the inherent problems that the bolection molds can add. Looking at cross sections, one can see that moldings are often twice as wide as deep. This will run the molding well out onto the glass, or into the raise of a panel. The panel needs to be able to move, and it is not considered good practice to cover up glass with wood. The solution is a landing, or land, that gives support to the underside of the molding and fixes the thickness of the opening. The lands can be made of a secondary species, and sized in a width to provide about a 1/2-in. rabbet for glass or wood panels. We would use a 1/2-in.-thick land plowed into the stiles and rails about 5/8-in. It should be snug enough to seat down into the bottom of the plow and maintain a good 90° from the edge of the door member.

The lands will give glue surface and mechanical support to the moldings, and will replace nailing as the primary (and often only) fastening. The lands will be made to follow any curves in the rails and/or stiles, but they can butt at the corners—no need for a miter or more. It is important that the lands have no gaps, especially at the corners, or the water may get in and begin the degrade. Once the lands are in place, the molding can be attached to the first side—generally the side to the weather.

Whether wood panel or glass unit, any bolection molding can create a good first defense against the weather. Reinforced miters will be what makes this defense work. A miter can be reinforced with a pin or two, a staple, nails, even screws. And glue, of course. But wood will make the best reinforcement. A simple wood spline bridging the joint will work to keep the joints tight and looking good as the next millennium approaches. The bolection molds can be mitered, glued, clamped, and then splined before setting into the door.

The miters can be stood on point vertically at 45°, and passed over the table saw or shaper to place a simple kerf across the joint. Size and drop in a spline made to run with the grain at 90° to the miter, and you will have a good miter joint for generations. Glue and clamp the bolection frame into the door. Use a nail or two if needed. Flip the door over, and clean up the miters and along the lands. If glass is going in, make sure it fits before drawing out the sealant beads. If it is stopping a wood panel in place, get trial fits all around, and glue and clamp the assembly in place.

The landed construction will also lend itself to very thick doors. A box-type assembled land was made for a 3-in.-thick door. This "box" landing gave support to the very wide bolection mold, and made for a stable platform. The miters were glued and clamped, and a spline was let in on the underside of the miter.

When setting a wood panel in a door with loose molding—bolection or otherwise—there is an opportunity to add a bead of sealant around the opening to gain an extra amount of insurance. I think it wise to capitalize on this if possible. In fact, with a 7-ply panel that is as dimensionally stable as plywood, it can be glued in place if we like. Now, that is a step I was not prepared to make when I first realized what I had could be mitered, glued, clamped, and then splined before setting into the door.

Doors that benefit from bolection molds are doors that like to get a little dressed up with a few curves, often in pairs. Indeed, the most complex doors we have made have benefited from glued bolection frames, but many more everyday doors have also utilized bolection moldings.

Chapter Six: Panels, Glass, Muntins, & Louvers

Panels

In frame-and-panel doors, the panels play a passive though important role: They fill in the gaps created by the structure that defines the door. They are called in-laid, inset, recessed, and more. Most people would think their origin of an aesthetic character, rather than a fundamental one. Indeed, they are often mimicked—and relegated to the pile of "decor" in lesser-quality work like MDF doors. We tend to call them "raised panels"—if they are, indeed, raised. We like to use the common blurred meanings that are prevalent in this craft. The word "raise" can be used as an adjective or a verb.

"Fielded" panels is another term, a bit British and slightly outdated, for panels that have a field created by cutting down the perimeter. The function of the thinner edges is to produce an edge that will fit into the frame. The decorative part came next, as it was realized this was a functional element that could add to the appearance of a frame assembly. The Shakers raised their panels all on one side, and turned the flat side of the panel out to be seen. This was done to avoid being seen as decorative or "worldly."

Passage doors evolved from large planks to swinging plank doors, then on to frame-and-panel in about the 15th century. The plank doors required rows of mortises on the edges of the planks that were filled with loose fox tenons that tightened up as the planks were driven towards each other. A massive amount of work to get an

improvement over a slab leaning over the hole. What event or thought caused the first woodworker to turn a plank 90° to the vertical planks and invent the rail has been lost to the centuries. But that simple act started a design tool that will never see its versatility and variation run out.

And as those stiles and rails run this way and that, we are tasked with filling the resulting holes they leave with panels. Panels are chiefly made of wood, with glass being a close second. Wood panels can be solid wood, or manmade board like plywood or MDF. Raised or flat are the two most common, with other solutions being rare and odd. "Raised out of the frame" is one I know of, and will discuss later.

Door panels can be raised with any of a nearly infinite range of raises: hip, ogee, ovolo, or beveled. They can be raised one side or both sides, or they can be flat or rabbeted with square or beveled steps. Very versatile.

The panels can be man-made boards, veneered hardwood plywood, solid wood, or 7-ply panels.

While the frame of the door does the work of holding a defined space, the panel does the major work of the door. Keeping the winter or the Cossacks out is the job that falls to the panels. The majority of door panels are wood, solid wood. Before the development of good wood glues, panels were one piece in width, or boards with tongue and groove panel boards to span the width.

I should mention that some manufacturers and some small shops place two panels in panel openings, back to back. Both are raised, and the strategy is to let one move to its environment (the interior side of the door), and the other can move according to its environment (the exterior side of the door). I have never felt the need to build this way, but there are those who say this solves the splitting panel problem. However, I have repaired split panels in doors built this way, so those making them may not be aware of problems. I just do not see two 1/4-in. tongues fitting into a 1/2-in. panel groove as being stronger than one 1/2-in. tongue.

Panels of solid wood are sized in thickness to fit into the groove made in the stiles and rails. Allowances must be made for the hygroscopic nature of wood, or the panel could expand and stress the stile and rail joints. If the door is to have exterior use, the panel must be tight in the groove so as to prevent driven water from making its way through the door. Panels leaking water is a major complaint with wood doors.

And like so many things in evolved door-making, there is no one absolute solution to any problem. The first part of our strategy to make a better door is also the easiest thing we can do to prevent leakage. We size/machine the panels to fit the groove very tightly. Thickness is what we want tight, but remember to leave room for expansion along the sides of the panel. The panels are to be sized correctly in width by learning the expansion likely for the panel and subtracting it from the opening width. The length of the panel is determined by the opening, with perhaps 1/32 in. less for safety in closing the joints. Any exterior door will have an epoxy end-grain coating, and that will take up that 32nd that we had.

The epoxy end grain coating—the second of our strategies to make a better panel—is a must for all solid wood panels. If water should breach the first line of defense—the sticking/panel joint—then we have water in the door, and it will seek out a path by capillary action, following the path of least resistance—the end grain of the panel. A coating of epoxy, preferably thinned about 50% with acetone, will penetrate into hardwoods as much as 1/4 in. and seal them effectively against any penetration. Epoxy is weakened only by ultraviolet light, so that coating on the panel ends will never go out of service. It will be working passively for the entire life of the door.

The third strategy takes us further along that design corridor in pursuit of a better panel, better door. We have the panel fit tightly to our plow first. Epoxy—end-coated and tight into the plows at the end grain second. The most serious problem we have ever had with our products concerns panel joint failure. We know that most of this was caused by defective glue that failed under heat encountered in most of the U.S. Other causes of panel joint failure are dull knives or bad jointing—panel joint failing, and not just a crack in the panel. As discussed, given the nature of wood, a panel can absorb moisture and try to expand in a frame with no room for expansion, and so will compress the fibers all along the width of the panel. Then, when the moisture content recedes, a crack may appear as the wood tries to adapt to this new dimension/moisture content.

Dull knives are easily remedied, though it sometimes seems difficult to interrupt the busy shop in order to set a sharp set of knives in the machines. The test for dull knives and the burnished wood surface they make is simple. Drop a single drop of clean water onto the freshly jointed wood surface. Surface tension inherent in the drop will make it form a familiar dome shape on the wood. In a few minutes, the drop should flatten and lose its shape as it soaks into the wood. If it does not flatten and soak into the wood, then the wood is burnished. Sharpen the knives and run the parts again, and compare the difference with the water drop test.

One other strategy for successful gluing for width is simple: Join (glue) the wood shortly after machining. Freshly cut wood will oxidize in a short time, and this lessens the wood's ability to take up an adhesive. I have not heard of this until just recently, but it seems like a logical connection. It is not always easy to glue immediately in even a small shop, but large shops, with hundreds of pieces being made every day, stacked and stockpiled, will never be able to take advantage of the clean, quick joint. I can not say I am aware of any failures caused by a several-day-old joint being put together, but I will accept the idea and add it to the strategies collected for good panel joints.

In some of the old manuals from when this work was taught in schools, a "hollow" joint is discussed. This is a joint that has both parts coming off the joiner with a slight gap in the middle of the length of the boards—1/16 in. if the boards are 12 ft. or longer, less for shorter stock.

The idea appears to ensure that the boards stay tight at the ends, with natural pressure augmenting the clamp pressure, keeping the joint tight long after the clamps are gone. I feel it is not necessary to hollow joint since the wood today is dried properly and to the right moisture content. A hollow joint might just induce enough stress to harm the joint in some way. Again, we want everything on our side.

We often get requests to build doors with one wide panel for the width of the door. These often go up to 42 in. wide and sometimes even 48 in. wide. This makes for a 30–38-in. panel—too wide for a "natural" width, so they must be glued. Now if probed, the customer often admits they think (hope) that a two-panel door will cost less than a four-or-more-panel door. While there are fewer parts, the size and difficulty of finding matching grains and then jointing such large pieces negates any labor savings.

These wide panels have given me some problems over the years. Mostly, I worry about cracks developing. And indeed a few cracks have developed. These are taken care of as warranty work, usually with a sliver of wood, some cyano glue, and finish touch-up. But, at times larger, more serious cracks may call for panel removal and replacement or "rip and rejoin"—which is to actually rip the joint open, joint both halves carefully, reglue flat and true, and add enough wood to make up for the ripped out wood.

As discussed in Chapter Two, glue failure has impacted our shop negatively. Additional costs, loss of work, frustration, and distrust are characteristics we try very hard to to eliminate from our day-to-day. Our living depends upon the strength and reliability of our joints. Panel joint failures early on encouraged a move towards more reliable joints, with reliable glue and reliable process. We had to do more, but what?

A first thought was to add a nice parallel grain or cross-grain spline to the center of the two butted pieces. The old rule of joint making—if wood is removed, replace it with other wood so no loss of strength is caused—is obeyed, but a good spline joint is not easy to make, and we had no spare machine space to dedicate to such a process. A proper spline joint has to fit tightly to the bottom of both plows, and also be snug enough to self-clamp in thickness, but not too snug so as to scrub glue from the wall of the plow as the spline is driven home. Too much glue may prevent the joint from closing, too little, and the joint is starved, with nothing gained. And there was no way of knowing what was going on inside that joint. It was truly blind—trust it and go.

And then there was the question: Would the spline, in either orientation, prevent panel splits in the future? We had no assurance that the spline would prevent the open cracks. If the glue failed, then it could easily fail down to the spline and maybe even further, gaining us nothing.

A number of finger-joint cutting tools for the shaper/molder are available, producing a sawtooth-type edge pattern that interlocks snugly. Finger-joints increase the glue surface area and present a great face, but the sawtooth pattern shows up in the end grain. Shorter "fingers" work better in long grain. Doing the math, I have found that some are better than others since they can actually double the glued area. That is a considerable gain, so well worth exploring. And then we realize. Most of our panels are solid, and raised. The sawtooth fingers will show up in the angled faces of the raises, looking odd. Smeared out, distorted, and "for no visible reason," the sawtooth pattern may not "raise" attractively. Finish could not be counted on to hide or disguise the joint.

A technique for panels on better work utilizes a manmade veneered panel with solid wood raised and mitered along the edges. The miters are sometimes splined or reinforced in some way. The raise stock is added to the manmade stable panels and mitered to fit precisely. The wood is flushed to level with the core, then the panel is veneered on both sides, and then raised conventionally. This is a time consuming process, though it will fit the requirements. Another advantage of this solution is that it eliminates the potential problem of end grain sitting in a panel groove with water. The side grain does not try to transport water like end grain does, but we have that one solved with the epoxy end seal.

Another variation looks more like shop-made plywood. A center core of man-made stable board will have a ply of solid wood laid onto each side, with the grain running the direction desired for the face of the panel. Then a cross-grain-thick veneer sheet—a cross band, 0.040–0.060 inches—is added to each of the two sides. Then a face-ply on each side of thick veneer, with grain and color matching from panel to panel and side to side. This makes a 7-ply panel, and it is now our current favorite in that it passes all tests with flying colors. Epoxy has meanwhile become the adhesive of choice. The resulting panel is stable, versatile, and allows for matched grain—continuous, book-matched, flitch-matched wood.

A variant on the above is a 5-ply panel—solid core from 1⅝–2⅛ in. thick, with cross-banding and then face veneers. When one examines a "5-ply architectural door"—a flush door—the core is man-made with two plies on each side. These doors are stable, and the assembly works, though there are no raises to contend with. We could of course make the center field of the panel and then add solids around the perimeter so they could be raised. We'll keep that one in the back pocket and bring it out only if needed.

Granted, this is an expensive endeavor—utilizing thick veneers, solid woods, resawing, grain-matching, joint-taping, vacuum-bagging, and epoxy. But seeing it in our doors makes it hard to turn away from. So this is the panel today. Subject to change.

The above solution also gives us the opportunity to match the grain from panel to panel in a door, as well as matching it in pairs also. This "side effect" of the process is serendipitous. One can never have enough serendipity in the shop, so the process is embraced and will be used going forward.

Another form of panel is tongue and groove boards, usually with a V-joint edge. The boards can be thick or thin, wide or narrow, random width or uniform. The patterns can include beaded ceiling, double V-joint, U-joints, shiplaps, and more. Most often, the panels are V-joint, 3/4-in. thick, and uniform width. We like to balance the joints so all panels have the same flat in width.

Some years ago, we noticed the standard manufacture on these doors included a 3/4-in.-wide plow sized to receive the 3/4-in. boards. However, this made a small, triangular void at the V-jointed boards where they fit into the rails. Unintentionally, this created a channel to direct water right into the heart of the door. Epoxy end-coating will help, but this will still allow too much water into the door. We tried some silicone sealant, packing it into the triangle, letting it cure, and then trimming. This was effective and better than nothing. But then we would see a board or two pulling away from the fill, leaving yet another gap. Something else would have to be found.

In problem solving, it has been learned over the years to not settle for the first solution. A mentor of mine, Bill Strode, was a man of few words, but with immense knowledge and ability. I was once in a true beginner quandary, not able to find a way to move forward on some project. Bill gave me his half-smile, and told me to find three ways, three methods, three processes to move forward with, and then to stop and pick the best. "Best" might be familiar, safest, fastest, or easiest, depending upon the solutions and the problem. This got me moving, and Bill watching, as I moved along to see if my choice was a good one. Bill would catch up with me in a day or two, and note which path I chose. Then the half-smile would come out, and he'd deftly explain two or three or four more ways I could have gone, none of which I had thought of. The depth of his knowledge was great, and his method at first was difficult, but the results are effective and still employed today.

With the panels, the first thought was to set a spline into the upper edge of the bottom rail, something about ¼ in. thick, with 1/2 in. or so projection. A matching kerf in the end of the V-jointed boards would fit snugly over the spline. But I did not like the 1/4-in. spline—I thought it too weak, and that a well placed kick could easily shear the spline and dislodge the panels. Use a strip of aluminum? That would work. Change the grain direction of the spline? That would work, but now are we too tedious?

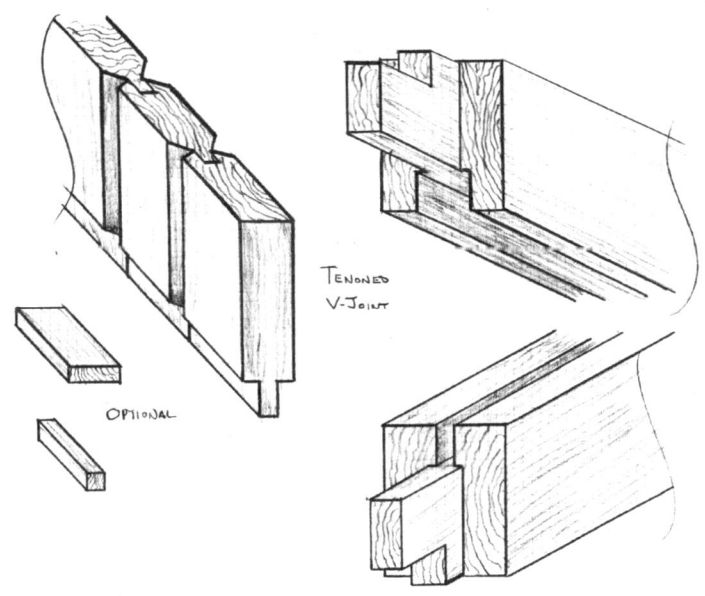

But most of what we do was once considered too tedious. Yet I will say I am a fan of simplicity. The reality, like so often, is somewhere in between, and shifting. I prefer simple until it is proven to not be good. I'll embrace complex if I have to—for the good of the product—only after trying for simple and elegant solutions. But whatever works best is the best in the long term, despite it once appearing too tedious or difficult.

Trying to stay with simplicity, I wanted to be able to work with what we had, not adding any additional parts or materials. Eventually the dim bulb grew brighter. Tenons. Tenons on the end of the boards. Simple. Easy to do with the tenoner. Can be done on the table saw. So we now have a good solid way to avoid water intrusion.

The lower edge of the upper rail and the stiles can be a 3/4-in. plow (by 9/16 in. deep) or whatever the panel boards are, with a slightly snug, rattle-proof fit. Only the rails with a panel groove facing up, or the upper edges of a cross-buck assembly need the special plow for the tenoned boards. A 3/4-in. board can make a good 1/4-in. tenon across the end, which is the same thickness as the tenon in the standard 3/4-in. V-joint. Thicker boards can have thicker tenons, with 1/3 in. being the guideline for ideal tenon thickness.

These boards are, of course, a panel, so they need some protection from water if an exterior door. The tenon groove (or "mortise") should be painted with epoxy—just enough to wick into the board ends and slop around between the rail and the panel.

But this glues the boards in! The alarm goes off. We all know better. First or second thing we learned—do not glue the panel in place! Well, sometimes we learn we can break the rules and why. We can glue these boards in place because they are V-jointed at the tongue and groove. This area presents very little contact along the edge of the two boards, so if they do expand (the most likely event in freshly dried lumber), they can crush along that thin edge and never need to move further. The crush zone will absorb the shrinkage and also allow the expansion.

Problem solved. Except…there are times when handling 10, 15, or 25 V-joint boards, glue, rails, stiles, etc. is just too much. Once assembled, if they all get pushed to one side, it could make the panel too narrow to fill the hole. The solution is two 1/4-in.-thick strips of wood, one just under the same width of the panel board's thickness and the length of the panel width, and the other the thickness of the board tenons. Put the panel boards together on the bench, exactly as you wish to see them in the door. Use a pinner or brads to nail through the 1/4-in.-thick strips into the ends of the panel boards. Two or more into each board will suffice—just enough hold to keep them from moving laterally. Make your allowances for expansion as you go—zero to 0.030 in. Now we have a one-piece panel, much easier to handle, and any clearances we deem necessary are already in place. This "panel" can also be glued in place since it has a crush zone intact and each board is still allowed to move within its own place, much like hardwood floors. The strips hold things in place to get them into the door at assembly, and also hold them in place for the glue to set and lock everything in where it needs to be.

Now, the problem is solved.

As the "casual" look in doors swept the industry a few years ago, the fake door industry changed their tooling to make the raised panels into a board-type panel. The raise was still there, the thickness was still there, but a series of grooves were added to each panel. Often a single clavos was placed at each end of the "board" as an attempt to further define the panel as boards. Simulated never works. There is no excuse for anything other than the real thing.

Another panel type, though it is not often seen, is raised "outside the frame." I have never encountered anything like an official name for this, so bear with me. Generally these are seen on 2¼-in.-thick doors, and the design is that of a very heavy door—as to a dungeon or worse. Instead of a large center plow, the doors will have a 1/2-in. by 1/2-in. plow just below either face. This allows a panel to be fit on either face of the door. The panel has a plow in its edge, so a tongue fits in the 1/2 in. by 1/2 in. plow that is 1/4 in. down from the face of the member.

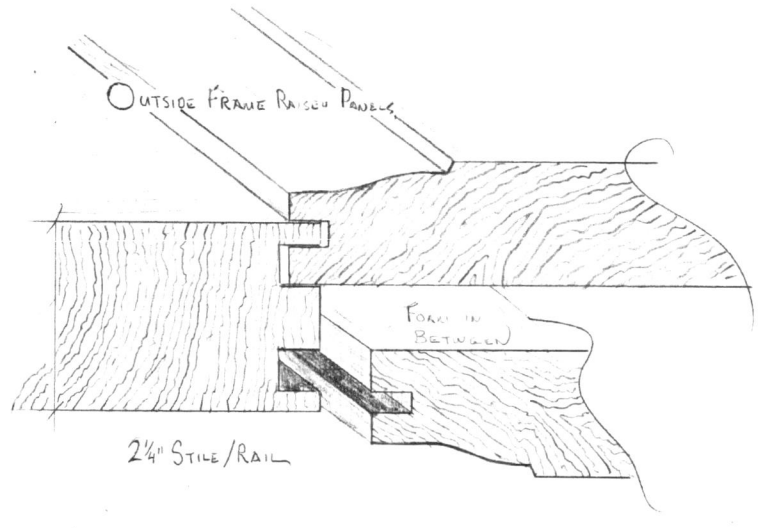

Panel thickness is dependent upon the panel raise and the desired remaining square edge. The panels are sized and raised and plowed. Note that if the panels were laminated, the exposed edge would need to be cleaned up to show solid wood so there is no complaint. The resulting door is similar to a flush door with planed panels. If you look, you realize there is no sticking, and the grain on the frame and panel behind give away the clues that the door is frame-and-panel, not a simple flush door dolled up a bit.

Acorn has recently developed a "new" panel type that we call a woven wood. The appearance is that the wood strips are woven, when they are actually more like tiles in execution. The woven wood panel can be made open with large gaps, or closed down to smaller gaps and even closed to a zero gap. This is a labor intensive option, but well worth the extra work. The finished product intrigues people once they see it, and it is counter intuitive when you can see both sides. These panels are reversible and work just like tiles. Chapter 12 has an extensive accounting of the development, the problem solving, and eventual solution to this interesting panel.

Glass

Adding a glass panel to your coped door is relatively easy. A square-edged door will have to be rabbeted for the glass, with either offset tenon shoulders to make up the difference or stop-rabbeted stiles and perhaps rails. Offset tenon shoulders are fussy, I have felt, so I avoid that solution if possible. While sometimes clumsy, stopped rabbet is the solution I am most likely to use should I make a square-edged sash with glass.

The coped door only needs to have its sticking removed for the glass to be fit after the door is assembled. This can efficiently and effectively be done with the simple rabbet cutter and bearing that is part of the tooling set, or a dedicated rabbet cutter, and the shaper fence can be used to remove the sticking where the glass goes. The objective is to gain access to the glass/panel plow from what will be the interior side of the door.

I like to have the sticking on both sides of the door at the rails for its contribution to glue surface, so if I wish to remove the sticking, it will have to be a stopped cut. The power feeder is swung up and out of the way, and a shop-made hold-down guard is made to guard the cutter, but allow the operator to swing the stile horizontally across the shaper table into the cut. The cut can be made this way since the material removed is small compared to the mass of the part being cut. With the machine running, the stile is roughly positioned longitudinally along the fence, tight to the fence at your back, but several inches from swinging into the cut. Marks on the stile indicate where the cuts should start and stop, and they help the operator position the part. Once correct, the stile can be swung firmly and slowly into the cut to full depth and fed forward. Do not try to "back up" or pull the stile back while in the cut. This will allow the cutter to get traction, and grab the wood and fire it out the in-feed direction. It will be gone before you can react. Where will your hands be? Any time wood is uncontrolled, it is dangerous, but uncontrolled and under power is doubly bad. Prevention is the only alternative.

Once the stile is in the cut, feed firmly and smoothly until the next mark is reached. It is not important to hit each mark perfectly; 1/4 in. over or under is OK. The hold-down/guard needs to be easily adjustable for tension so you can work without struggle. Feeding and swinging in the stile should all be easy, but still have that good amount of resistance that tells you all is well. It is good to have to push against some tension while feeding, but not too hard. Just enough to give feeling of solidity, a sense of security, that helps you easily maintain control. Do not fight a tight hold-down. It will wear you out, it serves no purpose, and it is not as safe as one properly adjusted. The stopped cuts are the challenging cuts, but due to the round cutter, the sticking is not fully removed square to the rail. That will be resolved after assembly. The rail cuts are simple and quick, a full cut the full length. These can be completed at the shaper on the same set-up as the stiles, with a full-length cut.

Dry fit the door to check fit and dimensions. This is a good time to measure the opening for the glass. Custom doors require custom glass. In my area of the U.S., the glass is almost always insulated, and from 1/2 to 3/4-in. thick, made to fit the door. The glass people like to have 1/8 in. clearance all around the glass panel, but I typically give it 1/8 in. per dimension, or 1/16 in. around the perimeter. It is not good to force glass into a bind (mallets do not work!) since it will cause the seals to fail early. A good fit is all we want.

The wood-to-glass joint needs to have a seal of some sort to keep water and condensation from making their way between the two. Today's woodworker is fortunate to have excellent choices at hand. Woodworkers have never had good seals, only marginal ones that were ineffective and often poisonous. Lead was a favorite ingredient in these compounds, later replaced in part by asbestos.

Today we use a room temperature vulcanizing (RTV) silicone caulk that is applied via a pressurized caulk gun. A 1/8-in. diameter bead is drawn on all four sides just inside the opening on the flat of the rabbet. Then the glass is carefully laid onto the bead, and wiggled a bit as it seats down, squeezing out some sealant into the opening, and also along the rabbet, making a complete seal. We have our door "face" down—the exterior side is down, so this first application of sealant is for the weather side of the door. Once the glass is set and centered, and the seal is complete, then a 1/8-in. bead is drawn on the glass for the interior stops to set into.

This second, interior seal is not usually done in the industry. Indeed, some makers forego even the first seal application as they try to appease the labor saving gods. They claim that the finish, whether paint or stain, should lap onto the glass and make an effective seal there. Any painter will protest, knowing the paint will shrink and either pull back or crack. This will allow water in, then the bugs, then rot. Very predictable. As an industry, doormakers should be ashamed of themselves, trying to hand off their responsibility onto the painters.

For centuries, door-makers tried to find better materials to seal between the wood and the glass. Water would get in and eventually foul the rabbets and rot the door. With no sealing solutions other than linseed oil compounds or white lead, the woodworker was at a loss, unable to solve the problem. But being a clever lot, woodworkers came up with a good woodworkerly solution.

The doors with glass were designed to hold a wood sash This sash could be repaired a time or two, and glass even replaced. The rot may start, but at some point, the entire sash would be replaced. This freed up the woodworker to make more interesting designs, and he could leave the glass unsealed and vulnerable, but now he could replace the sash.

The second seal effectively encapsulates the glass in a silicone wrap, reinforcing the seal the glass insulator has made. One glass fabricator, upon seeing how we seal his glass, voluntarily doubled his warranty (from seven years to 15 years). The RTV silicone also is an adhesive and bonds to the glass, will not allow it to slip or slide, and prevents earnest woodworkers from replacing damaged units. The proof of its effectiveness only becomes apparent when we have to remove a unit.

The wood stops for the glass should be made before the glass is set, and they should be fit—mitered—for a snug fit at all four corners. Press them carefully into place, into the sealant, until they seat nicely, and there is good squeeze-out all around. Pin the stops carefully or hand nail if that makes you feel less nervous. When I first was instructed to use a pneumatic staple right close to the glass, I was terrified. And I did hit my share of glass, shattering the panel, and making a mess to clean up. I decided I would hand nail from then on, using pre-drilled stops, and a hammer on a piece of cardboard to hit first the nail, then the nail set, as the nail is driven home. Slow, but safe, I kept that up until my confidence grew and I determined I would use the nailer and still survive.

In defense of the nailers, we use pins now, and only expect the pin to penetrate into the stile/rail by about 1/8 in. When nailing, one hand forces the stop down into the sealant, tight to the glass. The nail is then driven, so the nail is working in shear, not tension. Shear is where a nail has the most strength, hence the apparently short length. Once the RTV cures, in about 24 hours, the stops are in place for good, as the adhesive nature of the sealant holds them fast. We like to fill the nail holes with a solvent-based colored filler. This makes for a cleaner product, and avoids the painter's glob of wax after staining or painting. We have had glass companies replacing damaged glass call us to find out how to get the glass out—which side has the loose stops? They cannot find those poorly filled waxed holes, the sure sign of a nailed glass stop.

When we put glass into a door, we come under the scrutiny of U.S. building code. As a fabricator, we have the responsibility to keep our customers and end users safe. The federal government insists that we do a few things to that end. In certain situations where glass is most likely to break, or to cause serious injury if it were to break, we are to set safety glass in our doors, sidelights, and transoms. So, your work will need to comply with your local code—actually the code where the door is to be shipped because there are several versions of code books in use all over the U.S. Generally, the rule states that any glass below eight feet in height and within three feet of a door needs to be safety glass. Glass near a staircase or in a bathroom is also required to be safety glass.

Safety glass is glass that is treated in some way to prevent large shards of glass from making large cuts in people. Tempering is the most common treatment. It causes the glass to break into small pieces that are relatively harmless considering what annealed glass would do in the same situation. Tempered glass starts with annealed (or "regular") glass cut to size, with the edges ground a bit. This grinding helps release stresses put into the glass by the tempering process. Tempering consists of floating the glass on a lake of molten tin. This does indeed, make the glass harder and somewhat more impact resistant, though that is not the objective of tempering. Yet, while the glass is harder as a panel, it is softer on the surface than annealed glass and can scratch easily. Tempered glass is marked by a small, etched, logo from the tempering shop.

Insulated glass that needs to be code compliant will have both pieces—or leaves—of annealed glass tempered, and the finished unit marked with a logo. Imagine a pair of 15-light doors, each with glass units marked with the logo, roughly in the corners of the insulated units. Yes, it is objectionable. We always request any glass vendor to supply all code compliant glass as tempered, no logo. Some makers of insulated units will refuse, saying they are required to label each piece of glass with a logo, or risk violating federal law. However, there is an aesthetic exemption, as well as historic and art exemptions. You need to stand your ground, cite the code book if necessary, or find another vendor. In my market, there is only one glass insulator that will make small units, so we have learned to work together, and supply the best glass for the job.

Another form of safety glass is laminated glass. Laminated glass is made from two pieces of annealed glass, with a clear, flexible, plastic sheet between the two glass faces. This makes for a glass panel that, if broken, stays intact, holding the glass in the sheet shape, limiting the amount of injury caused. Laminated glass is heavier—twice the weight of annealed or tempered, but can be used in place of tempered glass. Insulated units made with laminated glass are heavier, but they also do not generally have the same insulation value since some of the insulation—the air gap—is decreased, lost to the added thickness of the glass unit. Why would one use laminated when tempered is available? Laminated glass has a greater sound dampening value than tempered units. Use it for glass in a children's playroom door to limit sound transmission. Or use it in a recording studio, set in foam gaskets on both sides, and tilt at a deflecting 5° for a professional solution to sound attenuation.

Insulated units with laminated glass lose some insulation value due to the decreased air gap in the units, but then gain some of it back with thicker glass panes. Laminated glass also does not scratch as easily as tempered glass. If you wish to preserve the air gap, then the glass unit will be thicker. A greater insulation factor is gained by designing thicker insulated units, but the aesthetics suffer since one side of the door is not the same as the other. It is an old story, form versus function. I generally err on the side of aesthetics, since that is visible from the start. I have never heard a client mention it was cold by their door.

There are limitations as to how thick insulated units should be, as driven by the sash thickness. With a 1¾-in.-thick door, and a 1/2-in. tenon set between two 5/8-in.-deep ovolo sticking patterns, we would ideally only put a 1/2-in.-thick (tempered) glass unit in the 1/2-in.-thick opening. This will allow the stops to be identical from inside to outside for the best look in the door. If we were to use two pieces of 1/4 in. laminated, with a 1/4-in. spacer, we are already 1/4 in. thicker than the tempered unit. This will require a different profile for the wood stops.

There are many variations of glass to consider besides tempered and annealed. We see beveled glass quite often, and they make it into tempered insulated units as well. Then there are many other types of glass: Low E II, beveled, German water glass, Kokomo seedy, French antique, circa 1910, and more.

Many are used as shower door partitions, so choose with care lest your door end up looking like it belongs in the shower. Any glass that is to be set by a professional shop must be code compliant. Homeowners can set whatever glass they like, as they are exempt from the law, but the recommendation is still there. We do not want to road test our liability policy, so we comply. There are exceptions for grandfathered/historic and art glass, but these are rare.

Introduce yourself to the local insulated glass shop. Set up a wholesale account and ask questions as they come up. A good glass vendor is required to be able to build doors. Shipping without glass is not a good option. Learn how your vendor likes to see glass ordered, and make no assumptions. Even if you are forced to get glass from some distance more than just around the corner, make the effort to be known in their shop, and make it easy for them to do business with you. As the little guy, you need to be compliant, willing, and able.

For about 20 years, all the so-called "mahogany" entries made in Malaysia with Indonesian woods had art glass panels sandwiched in the middle of the insulated, tempered glass. This art glass failed by releasing the "mud"—the sealant that holds the glass tight to the frame—as dust within the panel. This dust cannot be cleaned, and eventually contributes to the failure of the seals in the insulated perimeter. Observing this, we determined to find a better way.

We developed the "piggy-back" solution for art glass. This employs a rectangular wood stop to retain a typical insulated glass panel in place as normal. It is hand nailed and painted black to disappear in the assembly. The stop will be about 3/8 in. thick, and it creates a rabbet for the art glass to land upon, making a 3/8-in. air gap between the inside surface of the insulated glass and the art glass. Some art glass panels have reinforcement bars to stiffen the glass. These need to be accommodated by the size of the spacer, or turned to show on the finished installation. If the reinforcement bars are to go in the air gap, be sure they will not hit the glass when the door is slammed. A rattle in the door is never acceptable. Decorative stops go on over the art glass, and the package is complete. The art glass is to the inside of the building, so it is away from environmental pressures that can degrade the art glass panels, and yet is easy to get to if repairs are ever needed.

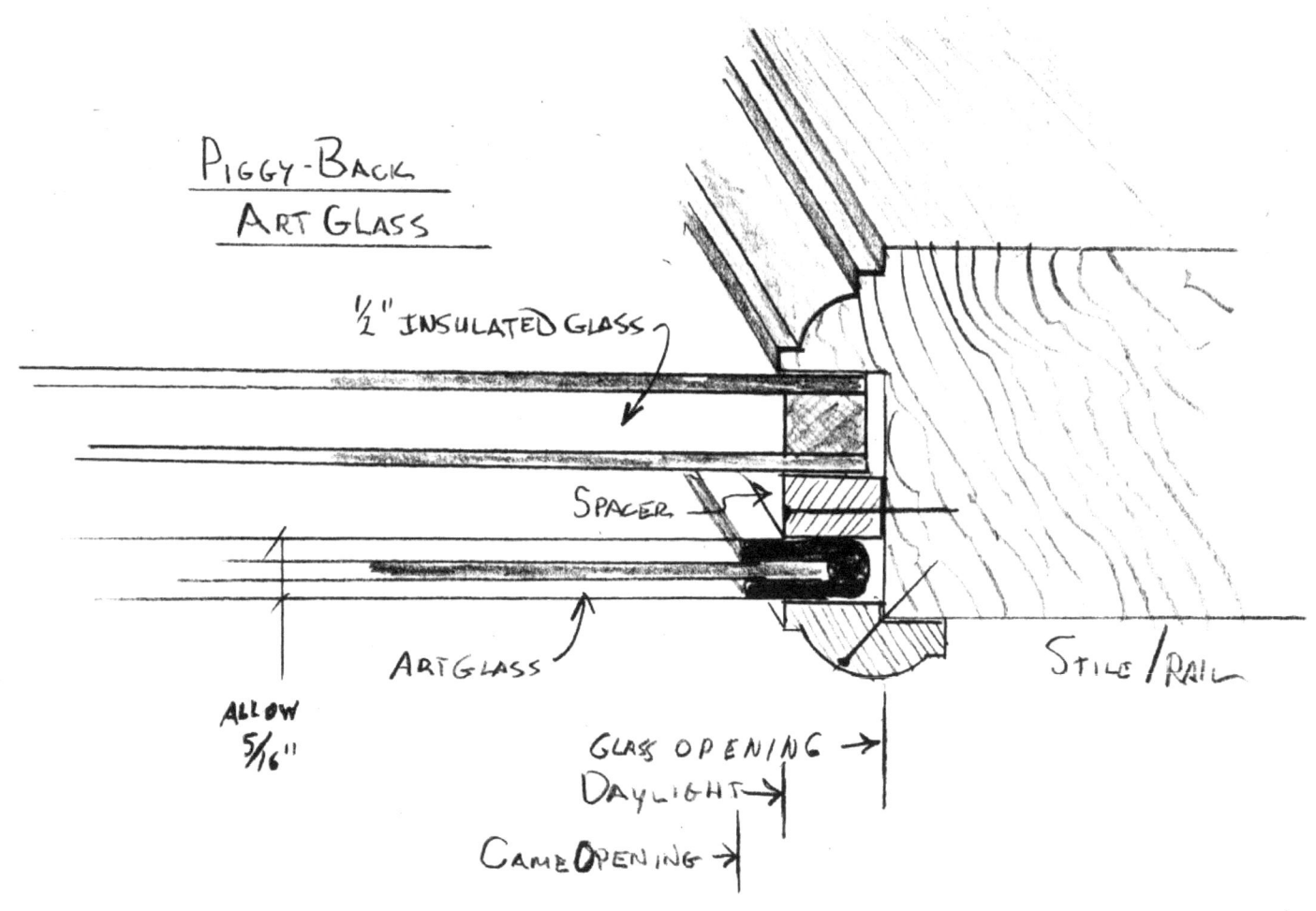

Muntins

Muntins, muntin bars, mullions, mutts, sash bars, and glazing bars are all regional names for essentially the same thing. The exception being "mullion," which is defined as a heavy vertical partition between sidelights and doors. In one shop, I learned that "mullions" was interchangeable with muntins, but the preferred term was "mutts." I learned how to make—these parts—in that shop, as well as most of the variants I have encountered.

When glass was a semi-precious material, and came only in small, fragile bits, it was important to be able to make sashes that could support a grid-work of muntin bars to support the glass that would then be glazed into those bars. Large pieces of glass were not available, and if they were, they would be prohibitively expensive, and extremely fragile. British taxes further increased the costs, so even the finest mirrors were made from several pieces of glass. Window sashes followed suit.

The resulting divided lights that are a familiar part of United States architectural heritage had their origin in those small lights. Linseed oil, whiting, bone meal, and a few other medieval sounding ingredients were compounded to make a glazing compound that would lightly adhere to the wood and glass to make a seal, and then to seal the exterior surface of the joint, hardening somewhat. This compound had to be regularly renewed by removal and replacement. This was done from the exterior of the building to keep the mess outside the building. Even into the 1950s, glazing windows was a skill that many had developed from setting glass in hardware stores or actual glazing shops.

Today, we have the freedom of glass of almost any size, and wood stops have replaced the comparatively temporary glazing compounds. Stops are placed from the inside of the building, so the integral stops are to the weather, making for a better, weather-tight sash all around.

Most of the muntin bars we make today are 1/4 in. "wide"—1/4 in. plus sticking two sides. The 9/16-in. ovolo will also gain 1/16 in. on each side for cleanup, so the total muntin net width is 1½ in., and final dimensions will be 1⅜ in. wide. They are made in the same way as the rails, but require special care and set-ups due to their narrow and fragile nature.

If a door has both horizontal and vertical muntins, the horizontal bar is the full width of the door opening, and the vertical bars run in-between the horizontal(s). Horizontals are cut listed at the same lengths as the rails, and will be coped and tenoned on each end, in a similar set-up as the rails, but with coping only on one face. The other face (interior side of the door) is rabbeted for glass, and will not have copes. The vertical muntin bars will usually only have a cope on one face and be square-edged on the inside face. No tenons, one cope, and one square edge. They do not require tenons, as the copes hold them in place, and they are clamped in place at assembly by a long clamp run vertically to seat the boards where needed. The horizontal muntins with tenons are tight to the stiles to secure the width, especially when the stiles are narrow.

The horizontals can be processed after the double-coped tenons are cut. The set-up stays the same (upper and lower copes on either side of the tenon) except one cope is removed, and a square cutter is substituted. The tenon remains. Then the set changes for the vertical bars, and the tenon is not needed. Size the bar in length as required, then cope and square-cut to length.

Once the muntin stock has all the end-work complete, it can be profiled and rabbeted. I think every accomplished woodworker I have met does this next step differently. All were successful and safe, with, I am sure some methods yielding better results than others. My method places the highest priorities on safety and accuracy. Profiling must, by definition, be accurate since the vertical bars will mate to the sticking profiled on the horizontals. I will mention that most people who became familiar with the tooling also developed their preferred way of cutting profiles and rabbets.

Feeding short muntin bars on the shaper will require a feeder, an auxiliary fence, and push sticks. While I like the ideal of the "endless board" as enabled by the use of the power feeder, I recommend muntin bars go through the machine one at a time. Profiling both sides of the bar, the exterior face will be relatively easy as long as the shaper has a good fence and the wood components are true and in good condition. End snipes are the most vexing problem we have with these today.

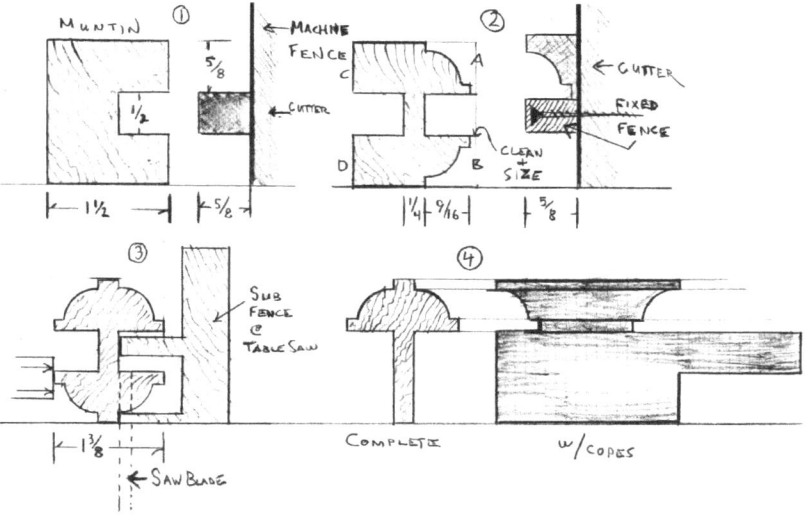

The sticking can be run with a 1/2-in. plow just below the profile, as part of the tooling set for the sticking. This establishes the surface that the glass will rest against. (See the drawing above.) These parts can be power-fed for the first profile cuts. If they trail away from the fence, they can be run again. Subsequent passes can be made with a shop-made hold-down/guard that will hold the parts down firmly, while a feather-board comes in from the side to hold firmly to the fence. This set-up avoids the threat of a tail snipe that can ruin parts. Parts cannot be overcut with this set-up, making for better moldings and better fits.

Another benefit of the process above is the last steps—ripping off the waste to create the glass rabbets produces a waste strip that is the exact length of the opening, or circle number. These can be saved and used as precise spacers to locate the horizontal muntins in place in the door at assembly. During glue-up, they are placed along the stiles as the top and bottom are brought together. Once the door is clamped tight, they can be removed and discarded.

Perhaps your experience would lead you towards the applied loose grilles that are prevalent in tract-house work everywhere. You may get calls from homeowners who would like to replace, repair, or rebuild their grilles since they are fragile and prone to damage. They are little devils to reproduce, but a fine money-maker if you can find someone to pay what it takes. Some window makers use a foam adhesive tape to fasten wood or plastic grilles to either side of a single insulated window, and add a metallic center spacer in the insulated unit to present a window that "looks almost as good as a real divided light window."

Custom wood door work depends upon having a market of potential buyers that wants what we do. When founding Acorn Woodworks, I determined to only make available divided light doors—or what the industry redundantly terms "true-divided" lights. I was proud of my ability to make coped-muntin-type divided-light work, and wanted to share it with anyone I could. The only rational reason for this was my own preference. Beyond that, it was historical precedent that appealed to me.

Then I read Stephen Mouzon's *Traditional Construction Patterns*. In discussing myriad reasons for divided lights, he writes: "the…play of light and shadow across the retina of the eye is physically more pleasurable than the harsher, more glaring light of an undivided window…divided light windows are inextricably linked in the minds of almost everyone with various human-based languages of architecture." In his estimation, anything else is a "cheap substitute" for the real thing.

So that was it—the simulated divided lights were just that: cheap substitutes. The simulacrum is never as good as the original. In fact, if one were to look at the removable grilles without knowing their heritage, one would wonder why such lengths, at such expense, were expended just to add some trim to a window. But the real thing is easily understood and bears examination as the true heritage of our woodworking and architectural history.

Louvers

Yet another panel type is the louver. Louvers seem to come and go in fashion. They peaked in the late 50s Early American wave, and have been low profile ever since. I spent quite some time in the mid 70s learning to make them in fixed and movable iterations. The old guy who taught me had been making them for 40 years at the same bench. He only had one eye (non-shop-related accident), and the joke was he could make louvers with the one good eye closed.

I would guess the louver has fallen from favor since no one understands "Why louvers?" Perfect for doors that need to cover HVAC, rest room doors, or where an air exchange is desired, but privacy can be maintained. There are two basic ways of making a louver door: Set the slats directly into the stiles, or make a sash that is then set into a door. As always, there are advantages and disadvantages to each method, so you get to weigh it out to see which path is best for your project. It is rare to see louvers specified from young design professionals since they just don't see them in their environment. So it is good if you have the trusty louver as a solution that you can pull out when needed.

When making louvers, start with a good cross section so you can accommodate any edge profiling, and plan the angles carefully. The repeat is determined at this time. Good practice with louvers dictates that there be a bit of overlap to keep the panels private and proper. A ¼-in. overlap would be a minimum, with 3/8 in. as generous. Plan the profile for the rails that meet the slats and continue the privacy. The rails will rabbet as needed, repeating the overlap and privacy.

There are three types of louver construction: routed, tenoned, and dadoed. Routed construction is by far the most common for commercially produced louvers. Stiles are routed and indexed to the next route.

Slats are fully captured, making this a good choice for exterior work. A jig can be built to hold and index the

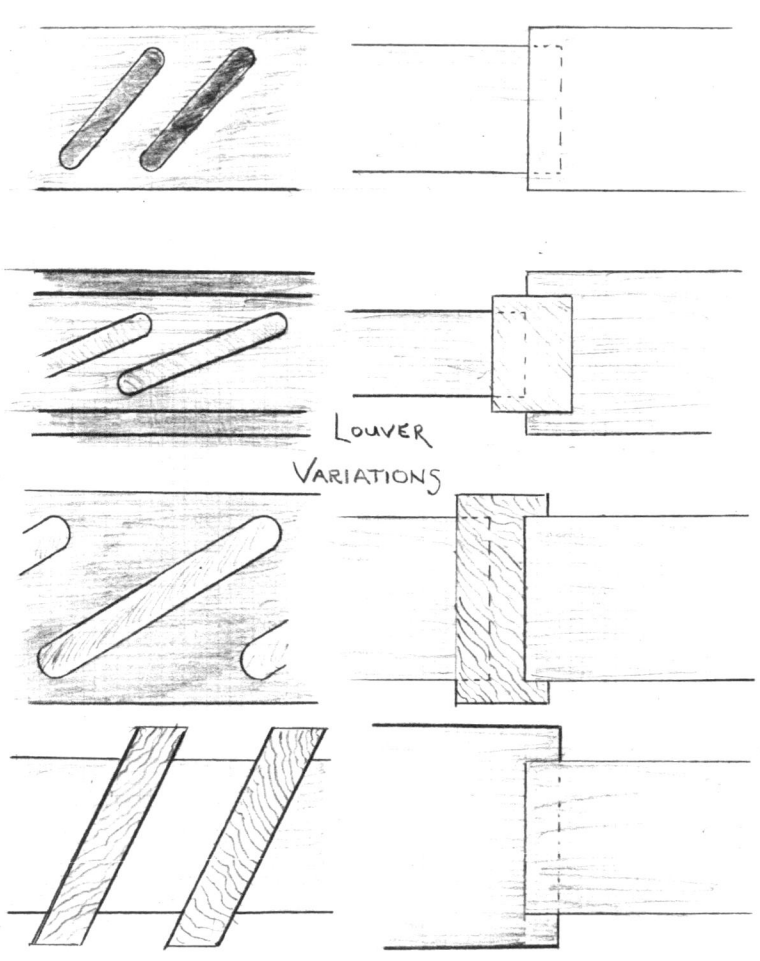

Louver Variations

two stiles. In better work, the rail edges are offset so as to "blind" the last and first slats. The routes should run about 3/8 in. deep, making all those mortises nearly structural. Draw these carefully, and do the repeated detail work to ensure things are on track. The angle and the overlap are important. Not enough overlap, and they are not secure (bath houses, changing rooms, etc.). A quarter inch is usually adequate. The angle should center the slat in the stile with a good 1/8 in. from the edge. Vary the slat width to get the dimensions you like. Wide slats are currently the fashion, but in the 1970s, fashion dictated 1⅛ in. wide.

Building a jig for cutting the slat mortises can be difficult to extremely difficult, depending upon your needs. The one I use, I built about 25 years ago, and it has served me well.

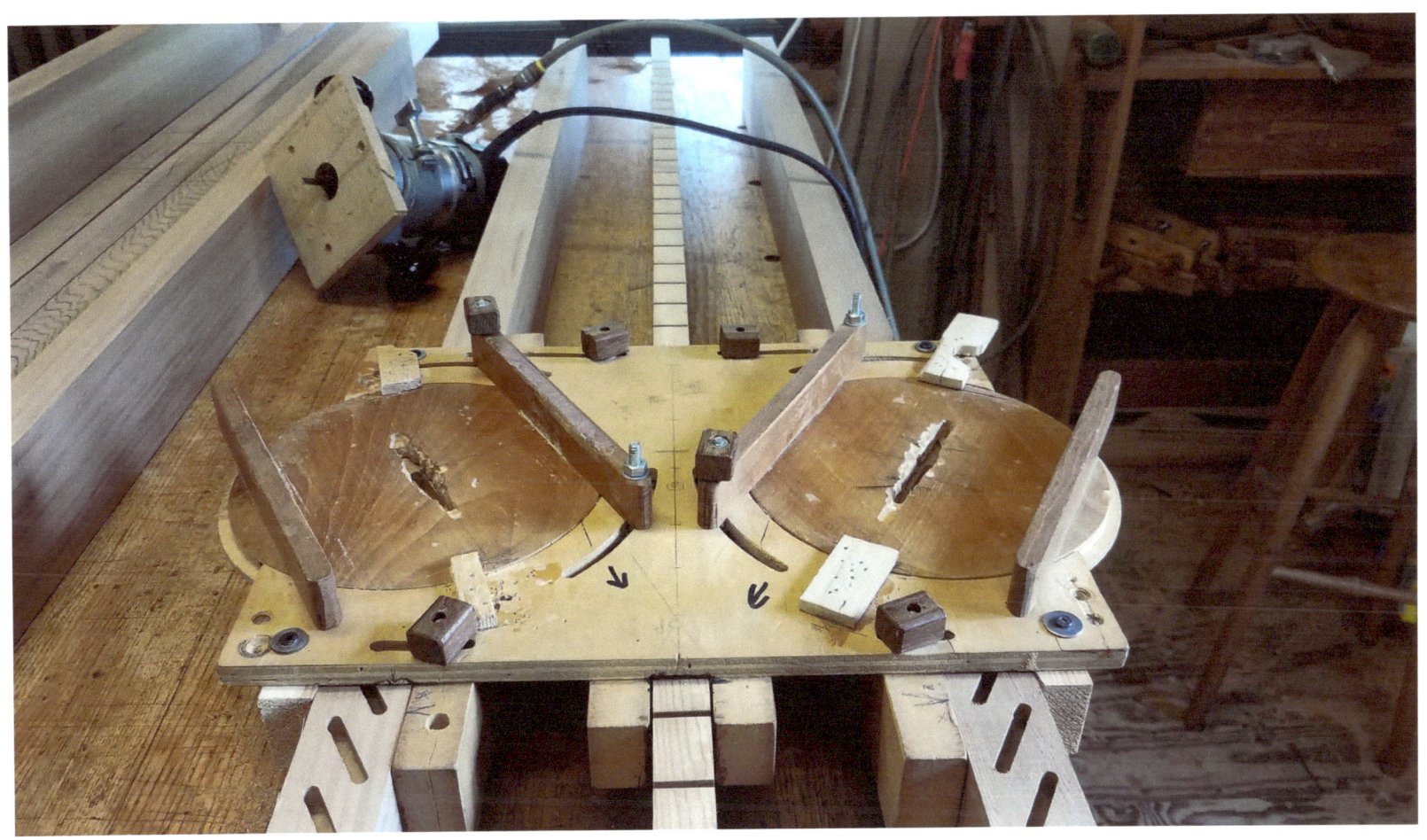

I can adjust the angle of the slat, the length of the slat mortises, the thickness of the slat, and more. I make an index bar for each job. It is simply a 1-in. by 3-in. by x-length stick, with 1/8-in. saw kerfs every 1–1¾ in., depending upon the slat spacing you are trying to achieve. Two long guides on the bottom of the jig fit on either side of the index bar. Four more guides run parallel to the index bar guides. These will center the stiles for routing, and will allow adjustment for various thicknesses of stiles. My jig can run 1⅜-in.-thick stiles on up to 2¼ in. or so. Thin stiles mean narrow slats and a strong angle, while thick stiles can be wider slats and an angle almost parallel to the face of the stiles.

The jig itself has two important parts. The first is the index tab. I used a 1/8-in. by 1-in. piece of steel and fixed it between the two router guides on the underside of the platform. This fits into the index bar easily, and will allow easy pick-up and relocation. It determines the spacing on all slats except curved head slats. The second important part is the router guides. My jig is built for a P-C midsize router, but I like the trim routers more for the lighter weight. I use a good, square base with the bit centered, and have it so the base is the guide for the router. If you are a big fan of collars for the router, they can work also, but are not as good at chip ejection.

The router guides on my jig are circles of plywood trapped on a framework of plywood, with clamps to hold the desired angle, and threaded parts to lock fences and to adjust the length of the mortise. I usually make one slat mortise a bit wider (.025-in.) than the other and fit the slats to it snug. A snug fit (as opposed to a "not too snug") is made on one stile, and the other side gets a looser fit. The fit is controlled by simple stops nailed with brads. The purpose of this dual fit comes at assembly.

All the straight slats are tapped into the tight stile, and can be held there as the panel is brought down and mated to the other stile. The tight slats can be worked back and forth to exit the mortise they are in, and begin to enter the opposite stile. Snug up the clamps a bit, and move on to the next one. It is good to work quickly since the glue is setting.

I fasten the stiles onto the bench with clearances set by my jig. Once set up, I like to turn on the router and do all the routes without setting the router down. Setting it down and picking it back up for each mortise, or pair of mortises, will triple the time it takes to route the stiles. Since many louver jobs will make 10, 20, or 30 panels, one can spend days picking up and setting down that router. The mortises are best at about 3/8 in. deep. This means the stiles can be run through the planer to remove 1/16 in. to clean up all the mortises and keep them clean and sharp. Over-width on the stiles helps when you want to clean up the router cuts.

Tenoned slats are typically ganged up and tenoned before ripping apart and rolling the edges. With slat thickness of 1/4 in., the hollow chisel mortiser is loaded with a 1/4-in. chisel, set at an angle to the stile. A 1/2-in. tenon length is good for this type of build, allowing the tenons to be nailed if so desired. Nailing a few tenons will help hold it all together. Non-corrosive nails should be used: brass, copper, stainless.

Dadoed slats usually project out from the stiles, or are sanded flush. Each slat must be nailed on each end to hold it in place. These slats will not contribute much to the strength of the sash, since they are not captured on all sides.

Moveable slats can also be made as a sash or set directly into door stiles. Specialized equipment is usually needed to nail the slats to the connecting pushrod correctly, but it can be done by any shop. The slats pivot on round end tenons about 1/4 in. in diameter, and 3/8 in. long. It is best if they bottom out in the bored hole, leaving a 0.020–0.030-in. gap between the end of the slat and the sash/stile for stain or paint. Moveable slats are a good solution to light control, but since they are outdated somewhat, they just are not considered.

Recently, we were asked to help solve a problem with wind in an outdoor area. Clients wanted to control wind when needed but open up when it was not windy. Our solution was to make a series of cedar panels that had horizontal slats, four inches wide on pivots. A pushrod and pins on one edge tied all the slats together and allowed easy control. Slats were bored for brass pins for the pivot, and more brass pins tied the end of the slat to the pushrod. Cap screws retained the pushrod in place on slats, top and bottom, and two midway. The pushrods were fit to a rotating stepped pivot that would hold the pushrod, and therefore the slats, at a chosen setting.

I have made side-mount pushrods for louvers in several applications. Most were more mechanical than decorative. But function rules with moveable slat louvers, old or new.

And there is one more variety of louvers—false louvers. Used for a secure door that needs to match louver doors, but cannot be a true louver. Or, for a purely decorative look, with no need to have them as true louvers. The false louvers are a double V, so the lower edge is down on both sides of the panel. A broad, short tongue and groove is made between the two "slats." The slats are run linearly for profile, and then cut to length and fastened to a perimeter frame that then is set into the frame of the door, usually as it is being assembled. Unlike real louvers, these slats are easy to lay out and make, and the angles and such are all easy to work with for the best fit and look. A spot or three of glue as they are going together, with minimal spacing at the perimeter frame fastenings, will allow seasonal movement.

Acorn Woodworks has a method of placing angled mortises in a curved rail to make arched top rail louver panels. These can be half-round, quarter-round or segment-arched. This will also work for angled heads without a curve. This is not a simple or easy solution, but it is a good, working solution that is far better than butting and pinning. I assume CNC can be programmed to do this, but it can't be easy.

The problem is an old one. In my research, I have found nothing that helps me see how it was done 100 years ago. Curved head louvers exist by the thousands—millions even? Half-round, quarter-round, elliptical, or segment-curve—they are all out there. And they all have louvers in them. I did see this done, once, in a shop in 1974, for a full circle louver for a gable vent.

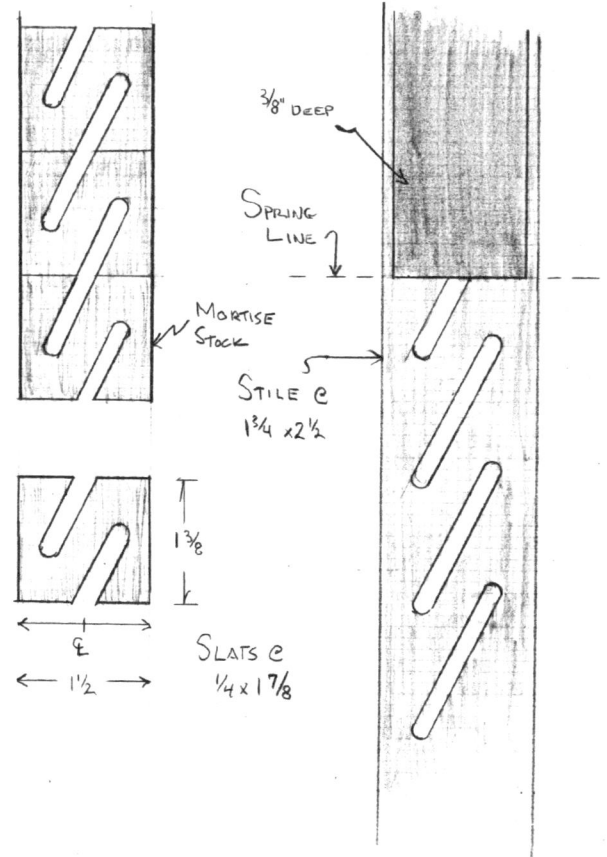

The straight stiles are made on dedicated machines today, from the ancient Festos on up to CNC. In a few minutes a pair of stiles can be made with ease. In our shop, it will take a few more minutes, but we can also make a fine pair of mortised stiles. The complexity comes when we have a curved rail with the slat mortises neatly in place. For our discussion here, we have a 90° curve, on a 14½-in. inside radius. How do we mortise this rail? Neither shop-made jig nor ancient Festo can do this. And program a CNC for the mortises? The people I asked about it conceded our methods just might be superior.

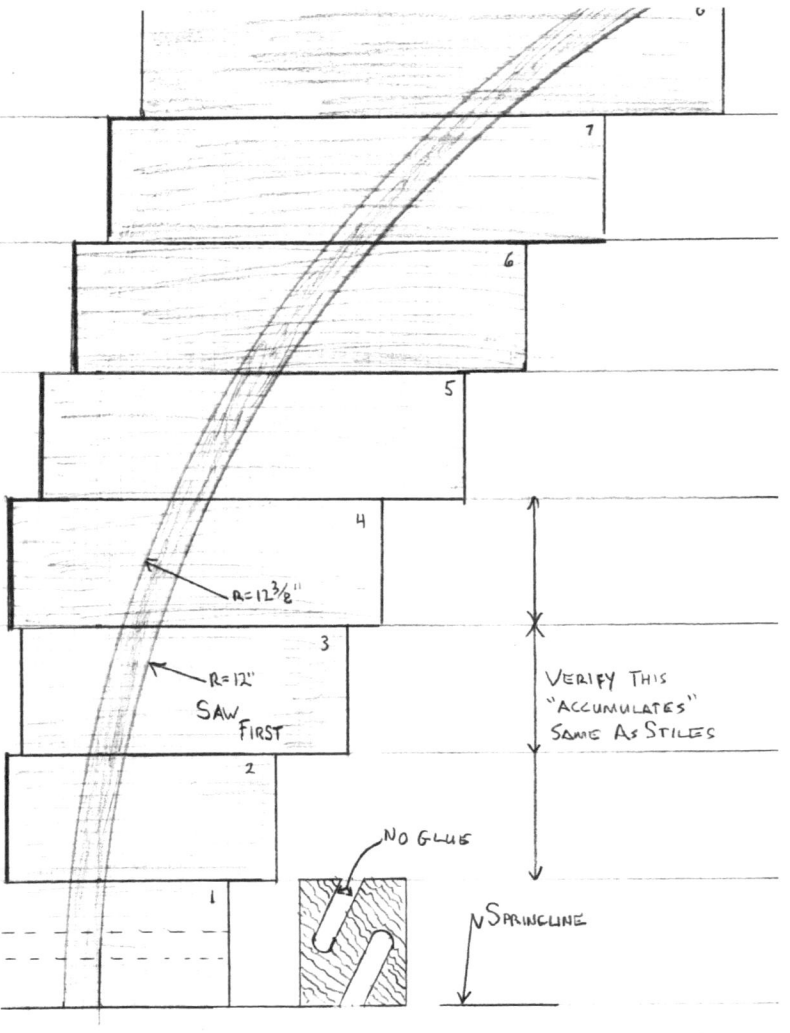

The method for this seems a bit arcane, as we head off in another direction from what seems to be the one we need. First, we draw it full size. Pay particular attention to the radii, labeling each carefully so it can be identified later. Determine the angle of the slat as it sits in the frame. Determine the thickness of the slat and the width.

We will use the table saw and dado saws to plow the thickness of the slat, or more precisely, the width of the straight slat mortises. Alternately, if you have a shaper cutter that is 1/4 in. thick, with a radiused end, and it can reach 1 in.—that would be ideal. The plow needs to be at the slat angle, and a bit (0.015 in. max) more than half the slat width. If you can come up with a way to bullnose the end of your dado saw cut, you get bonus points. I admit I use filler to close the gaps. I make them a tight fit, and that helps. I also turn the two 1/8-in. dado blades back-to-back so the bevel of the teeth make a bit of a point. Bullnose it ain't, but this works pretty well.

Returning to our shutter frame. Let's fill in some blanks: The thickness of the panels will be 1¾ in. The width of the panels is to be 17 in. The circle number (inside rail length) is 12 in. Therefore, the inside radius of the curved rail is 14½ in. The slats are ¼ in. by 1⅞ in. and sit at 26°. The quarter turn of the top rail will rise 14½ in. and run 12 in., and will require about 15 angled mortises. Make plenty of stock.

When you make the curved rails, be sure to add the ⅜ in. to the shoulder—the length—of the rails as they leave the spring line.

This looks like, and is, a secondary tenon. This will help hold the rail in place. Gain the stiles the same 3/8 in., and stop right at the spring line. If at all possible, make the spring line split on a slat, as in the drawings. After the rail is sawn and cleaned up, prepare to plow out the curve at 1½ in. high by 3/8 in. deep after the tenons are made.

Make the mortise stock next. About 25 L/F for each 90° rail. Make plenty of extra. That is a lot of extra footage, but you will need it. Once these pieces come out of the planer, stack them together and examine the total of 10 parts, and measure carefully. Compare to the measure of 10 mortises on the stiles. They need to be identical. Any difference here will show up in the panel as slats that are not parallel. A bit of a shifting out of parallel in the slats can be tolerated, but it can cause problems later on.

Be sure the 1/4-in. plow centers on the stock at the exact center of entrance/exit of the part. This ensures that when two parts are put together, the interior of the slat mortise is clean, with no steps. Be very careful making these parts, as any inconsistency can throw it all off.

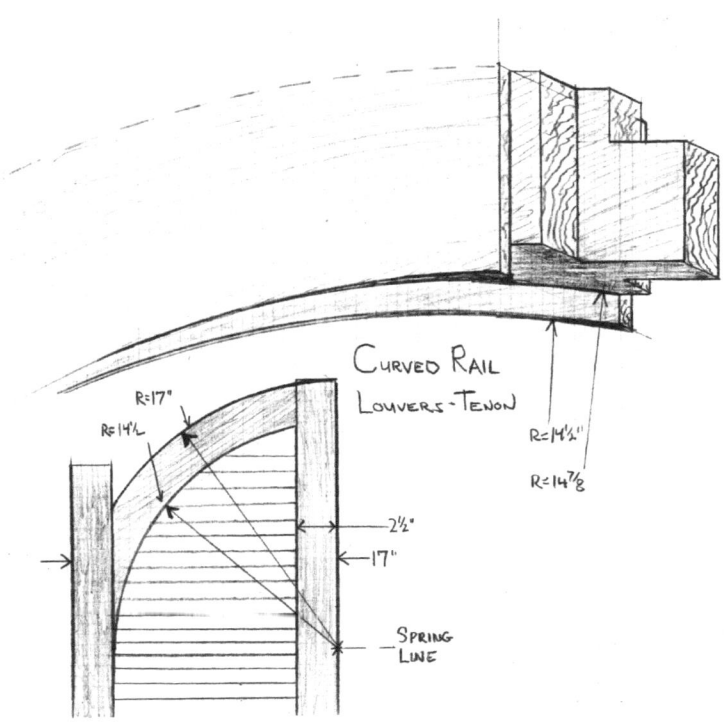

Cut the stock into parts from about 4 in. on up to 8 to 9 in. long. Lay them on the bench so that you can swing a radius on them with the trammel. Spread the glue on the parts sparingly, as we do not want the slat mortises to be fouled with glue. Clamp them up and let it cure.

Once out of the clamps, saw the 12-in. radius—the finished one—first. Clean it up by sanding the sawn face and scraping excess glue off the sides. Do not saw the second radius—12⅜ in.—just yet.

You can see the angled mortises as they develop, and you may have wondered how to cut the slats at the correct angle and curve. Take an assembly with the face cleaned up and the mortises ready, and tap a slat into each of the mortises, deep enough that the slat is past the 3/8-in. depth. Then go to the bandsaw and saw the 12⅜-in. line. This will automatically put the odd angle and curve on the end of each of the slats, while at the same time freeing up the slat mortises that will slip into the plow in the curved rail. Leave the slats in the rail so you can measure out and mark them all where they need the other end cut for the regular mortise. I usually saw these in the rail and remove the line to give them a bit of clearance.

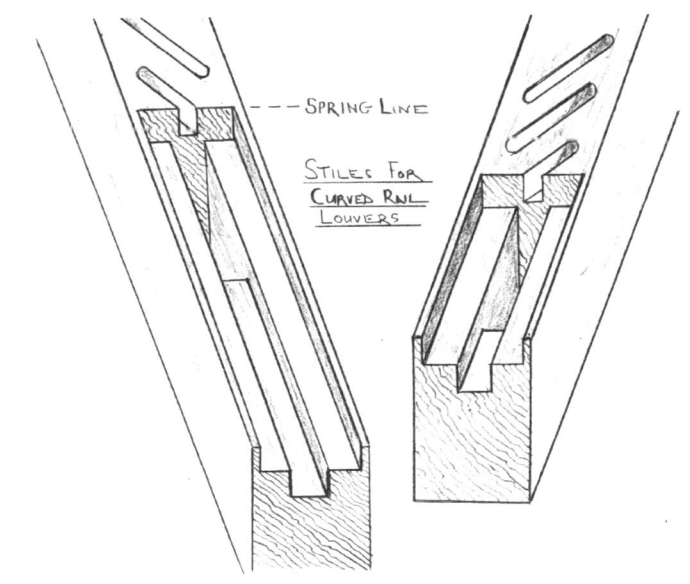

Remove the slats once they are cut to length. Do not intermingle with slats from the other side of the curve. The 3/8-in.-thick curved slat mortise assembly can now be glued into the rail, taking great care to align the slat mortises with the straight stiles to prevent an out-of-parallel situation. The curved rail mortise part can be slid along the plow to adjust for parallel if needed.

Prepare each stile to receive the rails, both curved and straight. The 3/8-in.-deep mortise is best if you can end it on a center of the slat mortise. You will not have any control of the landing of the upper slat/mortise. If that last slat is just a sliver, then it might be wise to slide the curved mortise assembly up or down a bit to remedy that situation. Do not slide more than 1/2 in. or so, as it throws the geometry off and the slats will not be parallel.

When you get to the panel assembly, worry the curved end slats into place first, and then fit in the straight ends. Keep the curved end slats a bit loose, as they will tighten up as they go together. Clamping and tapping with a mallet will help get it all together. The panels will require a bit of knocking around to get the stiles straight and to insure all the slats are fully seated.

Louvers can be combined with flat or raised panels for a very custom look in shutters. For a period, it was fashionable to make a cut-out in the panel— a single pine tree, a five pointed star, a quarter moon.

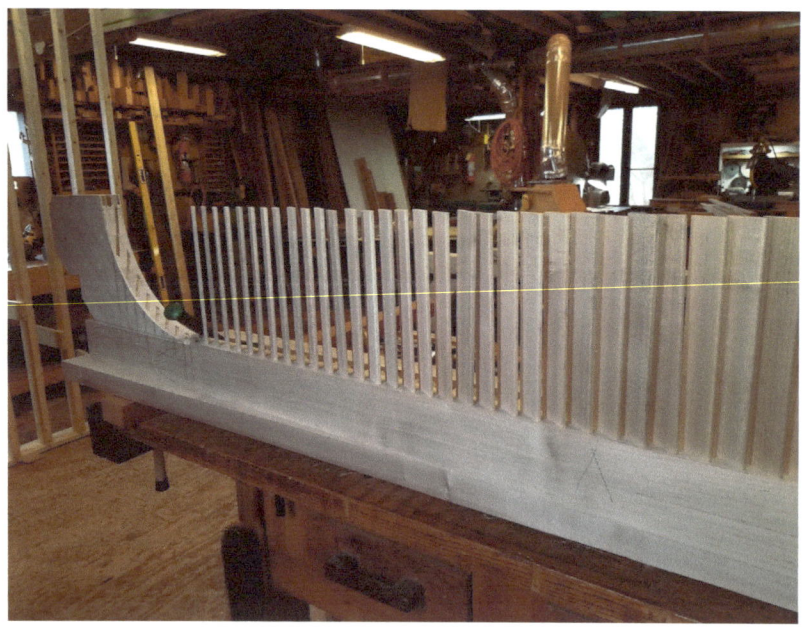

Chapter Seven: Plank, Flush, Secondary, Dutch, Radius Plan, Compound, & Large Doors

Plank Doors

A plank door is an earlier door form than frame-and-panel doors. The original doors were made of vertical planks of equal or unequal widths. These were joined edge to edge with mortises made along each plank, and loose tenons fit, often with a fox wedge to tighten up the joint as it closed. These doors were of heavy boards, and made for heavy use—castles, drawbridges, cathedrals, fortresses, and other uses, often with security as the major motive. Hingeing and latching were also monumental problems, solved on a case-by-case basis by enterprising souls on the spot.

As time moved on, and tools and the men who made them got better, doors evolved inlaid horizontal and diagonal braces to keep the planks aligned and prevent sagging. This helped forward the design and use of more effective and precise locks and latches. The bracing was inlaid into the thickness of the planks on the inside surface of the door, and was "tree-nailed" into the planks with short dowels hammered in. As metal became more available, nails were driven through all the members and clinched. When the peasants would gather at the church, waiting for those massive doors to open, they would often twist and pick at the clenched nails, hoping to get one or more loose ones out to take home as a souvenir. So, the smith made nails of a larger diameter, and formed a head on one side, and a washer for the other, and clenched the heavier nail on the washer, with heat, to prevent the petty pilferage.

Recently, we have seen an increase of the use of what are now called "clavos"—Spanish for nails. Designers have discovered they can quickly help move a design towards the rustic. Granted, the clavos were placed arbitrarily, and not located as functional parts of the door. The original nails pierced the joints and helped lock the mortises and the lap of boards. Again, we know that simulated has no place in good work, but when the customer requests...we oblige. Thankfully, this fad is fading.

But the basic look of the plank door has survived well over 600 years, and goes well with older designs of a mostly European nature. Well before the 20th century, probably in the mid-1800s, plank doors evolved a bit as they moved to thinner boards and more efficient bracing. The boards were either grooved on each edge with a loose spline inserted, or they were fit with a tongue and a groove to align and close the joints along the boards. The horizontal boards braced the planks and held them together, using the more prevalent nails to hold it all together. Two horizontal boards with one diagonal became a reliable and easy to build solution.

The doors were of a stable enough width to allow a closer fit to the jamb—by this time wood, or maybe still masonry. The closer fit enabled better hingeing and latching—and resulting security. If increased cold repellence or security was needed, the door-makers added a second layer of planks over the bracing, achieving better protection, and greater warmth. Doors like this were easier to build compared to the large, heavy plank style, and buildings of all sorts now required doors to keep people and weather in or out, and things in place in general. With the improved stability of doors, locks advanced, and hinges developed, making the clever smith and capable woodworker valuable assets to a community. Locks moved from furniture to doors, and entire rooms could be secured with one locked door.

Through its history, the plank door has had mortise and tenon construction at heart, though that is not obvious until we consider its history. Today, new plank doors honor this historical heritage in what we call a ladder core for the door. This core is a mortised and tenoned assembly in the center of the thickness of the door that will hold the door shape flat, square, and true. The planks or boards are fastened to the plank door by various methods, and they provide the look desired with all the functions of any modern frame-and-panel door.

Speakeasy windows, with or without doors, often graced the popular plank doors in the Midwest early in the 20th century when Tudor Revival houses were enjoying a resurgence. These smallish windows often had a bit of art glass fit; maybe a family crest, or early logo. In the then-evolving suburbs, the small window gave an element of privacy, as the door did not have to be opened for just anyone.

White oak was popular at the time for these doors, with most boards of random width, and of a quarter cut to lessen expansion and contraction across the width. The movement of the wood across grain was limited in the same way a group of boards on a floor had a limit to their movement. Each floorboard is nailed in its own place, with a small gap to the next floorboard. Today's kiln-dried lumber makes this much easier, since we have predictable moisture content in the lumber, and in the wood object's service environment. We will build our plank doors with an expansion gap to allow the boards to move within their own space. As we add planks to the ladder core, we use 0.025-in. plastic spacers to hold the tongue and groove boards apart.

When we buy kiln-dried lumber, it arrives as dry as it will ever be in most cases—6%–8% moisture content. This means the lumber will gain some moisture, increasing its dimensions accordingly. This is true for both sides of the door—one side being interior, it will move towards the equilibrium moisture content (EMC) for my area: 8%. The exterior of the door will reach about 10%–12% depending upon exposure. The ladder core's additional function is to hold the door at a consistent width (and length) so it can be fit to an opening and function as well as a frame-and-panel door. The planks at either edge of the door are glued to the ladder core stiles all along their length, but only about half their width—from the outer edge in. Additional planks are glued in place at the top and bottom, and at each "rung" of the ladder—or horizontal rail. The spacers hold a small gap for expansion, allowing each plank to expand in its place should the increased MC require that mechanical reprieve.

We build ladder cores with the same species as the planks, with several cross rails—maybe six or seven, depending upon the height of the door and the spacing apart. Stiles need only be 3–4 in. wide, unless a mortise latch is to be fit. The ladder core is often only 3/4 in. thick, with 1/4-in. tenons in 1¾-in. deep mortises. Additional frame members will be needed if there is to be a window. Once assembled and cured, the frame is extremely rigid.

Adding tongue and groove boards to the ladder core is easy. They can be clamped only, or pinned in place, and then clamped. Align one long plank edge flush, with the boards long, overhanging a ladder core that is also longer than needed. Work across the face, gluing, clamping and curing according to the glue, and be sure the last board has enough width to project a bit wider than needed. This will be jointed or sawn off once the planks are fit to the other side. The entire side can be laid on, pinned to hold it in place, and then slid into a vacuum bag to press the planks in shape. Turn the assembly over and cut rigid foam insulation to fill the voids between the ladder rails. The foam can be sent through the planer to adjust thickness, and then the table saw for the other two dimensions. Make it a close, firm fit to cancel drafts and driven rain.

Plank doors work best at 2¼ in. thick since that allows us to build a 3/4-in. ladder core, and then have two 3/4-in.

tongue and groove faces. Our tongue and groove cutter set is actually a V-joint cutter also, but it allows us to make tongue/grooves as small as 3/16 in. thick with a 1/4-in. face to one side of the tongue/groove and a 1/8 in. under the tongue/groove for planks at 9/16 in. thick. A 5/8-in.-thick ladder core, with 1/4-in. tenons and the two 9/16-in.-thick faces will make for a 1¾-in.-thick door.

As the ladder core is readied for boards, lay out the the boards to insure they will cover as anticipated. Use the spacers (these are 0.025 in.). Then spread glue and add the first board. Add subsequent boards as the glue sets up, working across the face. Once the first side is planked, turn the door over and add rigid foam to the interior. Foam can be run through the planer and the sander to thickness it to the size required.

As mentioned, the planks are tongue and grooved, but they are also V-jointed, with a 45° bevel slanting down to the tongue or the groove all along the length of the boards. This thinned edge comes together and will allow the wood to crush a bit upon expansion. This crush zone prevents buckling as the wood gains moisture. All wood hits its equilibrium moisture content in service according to each unique environment. Once the wood goes through a few seasons of movement, it will not register any more movement of consequence. A marvelous and very quiet bit of engineering, the crush zone is where we can find some forgiveness. As we know, we all can do with a little forgiveness.

Add boards to the second side just as the first, using spacers and clamping as you go. Once the last board is set up, you can trim the width, bottom, and head of the door. The head of the door in the photos is a two-arched Gothic arch. A template was made for the radius of the arches, and it is located on the head of the door. A router is used to trim flush to the template. This door used pivot hinges and a hydraulic floor-mounted closer, and is set up for commercial use.

An alternate method of adding the boards is to make the boards for each side a single panel. Cut the boards to the same length. Use a ripping that is the same thickness as the boards, and about 1 in. wide. Staple or pin this onto one end of the boards, and same at the other end. Use your spacers when fastening the cleat onto the panel so your clearances, as well as door width, are maintained. Then you can take the three pieces—face assembly, core, and second face assembly—and slide them into the vacuum bag and let it do the work. Twelve hours later, your door is ready to remove from the bag, sand, add hardware to, and hang.

Plank doors are versatile—they can be very tall, very wide, and even thicker. They can have one or more windows, and the windows can have glass, wire, or doors over them. Even widths, random widths, horizontal and/or top rails, inlaid diagonal braces…the variations are endless.

Flush Doors

Modern flush doors came into style in the 30s with advances in veneer-slicing machinery and the post-Victorian era of early minimalism and Arts and Crafts. Simple was shown as the strong rebuttal of the clutter of Victoriana. And the world of veneers was in a golden age of sorts, with thousands of fabulous flitches of exotic veneers from all around the world. As hide glue became more of a predictable commodity, and other adhesive options began to make their way into the market, veneering became a viable and valuable skill that could extend limited resources (though that was hardly a consideration at the time) and provide backdrops for the design and social change that would mark the time between the wars—WWI and WWII, that is.

The 1880s saw advances in veneer-slicing that made for an onslaught of veneer. Quartered white oak was the most desired of the veneers, so it was sliced by the acre to find its way into furniture up until the 1920s when the Great Depression slowed production of core goods.

This flurry of veneered goods became known collectively as "golden oak," and it dominated interior design for decades. However, the material became a sort of wallpaper, applied to anything that previously had been fine as maple or cherry. So secondary woods became primary woods, by virtue of a thin layer of veneer. "Tacky" was the appellation granted to this lot of work. A sort of industrial revolution with little or no restraint, and uneven supply and execution. Once the corner of veneer lifted and exposed the hide-glue-covered substrate, tacky was the best you could say. Humid weather exacerbated this lack of adhesion, and summers were often followed by re-tacking veneer in the fall.

Today, we have many good choices for adhering wood veneer to a core with little or no problem. But then, this is wood, and we have had our share of difficulties. As a woodworker, I am often humbled, and I will admit I have learned a lot about humility when I work with veneers. A slight bubble, a corner lifting, an open joint that was once tight: all lessons in humility.

Balance. The door needs to be in balance. If 7-ply, there is the core, then three layers either side—the face, the cross band, then a thin core, or the faces could be made up of a face veneer, then a cross band, both placed onto 1/4-in. ply. Two of these can make the faces for the door when placed correctly. I think it best to have cross bands in contact with the face for the best stability. Any method that keeps the door in balance will be respected.

We have stayed away from processed veneers—paper-backed and 2-ply are two popular forms. And we rarely buy the veneers laid up. Stile-and-rail wall panels would be made with outsourced laid-up panels, as quantity demands. Regarding outsourcing in general, we have found that no one cares about our work as much as we do, so we might as well do it. Admittedly, the number one problem/fear with veneer is sanding through to the substrate. Processed veneers of any type are thinner (0.015 in.) than any raw veneer, which is between 0.035 in. and 0.040 in. Sand-throughs are nearly impossible to repair, so just one sand-through can ruin a panel. If it is part of a matched sequence of panels, the sand-through can be catastrophic.

The raw veneer comes from several sources. We prefer thicker, for the obvious reasons, so we look for stock that is 0.0625 in. (1/16 in.) or thicker. This is a good thickness to work with; it handles almost like solid wood and can be stacked and sawn with a good track saw or edged on the joiner. Masking tape will work for joining sheets, but veneer tape is better when it has to be removed. We have found that a zig-zag stitcher is superior for edging sheets together. The heated glue thread holds it all together while being so light and flexible that the thread can go glue-side-down with no problems. This is much faster and more reliable than veneer tape.

The flush door itself is typically a 5- to 7-ply construction, balanced, with equal materials, species, and thickness on both sides of the door. A strong effort at balanced construction is made to ensure a flat door is the result of our efforts. The core is what we call an "egg crate" for its construction from sticks. The sticks are carefully S4S'd and half-lapped together to make the core. A thickness of 3/8 in. to 5/8 in. is good for each stick, with the width being driven by the net thickness of the door. The half-laps should be snug, but should have no chance of bottoming out before the two sticks level out. A bit of glue on the four faces that will touch each other is helpful in making the egg crate flat. The faces should be made on a bench that is flat to help with the "start flat, stay flat" mantra.

Lay out the half-laps to make rectangles no larger than 7 in. by 7 in. We like a 1/4-in. MDF face on both sides of the egg crate, and smaller than 7 in. may allow the 1/4 in. to sag into the cells under maximum vacuum pressure. If the door is an exterior door, we would fill the cells with rigid foam, planed to fit flush to the egg crate. Rigid foam in each cell will prevent sagging under pressure, so the cells can be made larger than 7 in., but we stick with the 7-in. guide since it works. We then prepare a layer of cross bands for each face of the door, and then the faces.

The sticks that make the egg crate can be processed easily with a bit of tape. Gang up all or most of the sticks that will run vertically, tape them into a wide, tight bundle with tape in about five places. Orient them so the dado can cut across the width easily. Set the depth and make the cuts. It is not necessary to make a precise layout on the sticks. Just a casual marking and cutting will suffice; just make a witness cut or pencil line so you know to keep everything going in the same direction, so to speak. Do the same with the horizontal (shorter) sticks—but you may need to make two bundles. If so, be a bit more careful with your layout marks and ensuing cuts.

Again, a witness mark will help make for uniform spacing and a better panel.

Add a drop of glue at each intersection and lay your sticks up to make the panel. Be sure they all seat down and are level. The vacuum bag, on a flat bench, is good for this. When sizing the dadoes, do not make them too tight or too loose, otherwise there is a fight of some sort. Let it be flat of its own will; the skins will adhere better, and the result will be good.

The door edges need to be dealt with earlier rather than later. We can add a single 1/2-in.-thick x door thickness stick of matching/appropriate species to each long edge of our egg crate to get the edges to match the faces in species. Blocking can fill any cell that might get hardware. But in order to get the edges to look right—just the species that is required, with nothing else showing, the sequence must be considered. We would add the 1/4-in. MDF to each side of the crate with yellow glue, and flush it to the egg crate sides with a router. Then we add the matching species to the sides. These are leveled with planes or carefully with a sander. Then the whole assembly goes into the vacuum bag for pressing with cross bands and faces added for the press.

We use epoxy for veneering. While it might seem to be overkill, we have encountered problems with every other glue made for veneering or otherwise. The veneer faces in the doors and panels that are epoxied come out of the press with about 50% of the face showing where epoxy came through onto the face. This will sand off, of course, and it does not show when finished. Epoxy does fill the grain where it bleeds through, especially on open grain woods. The finisher should be made aware of this since it might prevent some finishes that use colored grain fill.

The finished, flush door should be marked for top and bottom, especially if there is latch blocking in the door. The core/egg crate helps keep the weight down while holding the door flat. And the faces are stable, smooth, and flat-surface-faced with veneer as selected. We have made doors without the cross bands, but we had some minor problems with seams showing. So we will just swallow hard and move on. Fortunately, the cross band material does not have to have AAA premium veneers. It does not even have to be the same species in order to do its job—just taped with tight edges and glued like the faces, with epoxy, rolled on with a 6-in. roller. We are not gamblers, so we will create the sure thing, and there will be no question.

Once the faces are sanded on the benches, we also will sand the edges, and then put a slight chamfer on the veneer edge to prevent damage, and then add bumpers to the bottom of the door for harmless handling. Give the door the eye to see if there is a crown, and mark the top rail with some sort of mark to denote if the "belly" of the door is this side or that. If the door needs to have a crown, that should happen just before one side is glued in place. We would use veneer scraps to support the edge of the door along its full length, adding a shorter length, and then another, to create a crown at the middle of the height of the door. The crown should be no more than 3/16 in., and the veneer layers can be taped to the door face so they will influence the "flatness" of the door at each subsequent step, adding another layer. The reverse should happen on the other (second) side, so when it is pressed, the curve is formed. The seven layers, all glued with the crowning veneers in place, will hold the shape forever. Crowning will make the door tight to the stops/weatherstrip at the top and bottom, contacting there first. As the door is closed, once the bolt is engaged, the door will be tight all along the stops, and the latch will have a bit of tension on it, eliminating the possibility of rattling.

There has been a fad of sorts that has passed through the shop in the last couple of years. (That is the advantage of observing for 50 years—it is easy to spot the fads as they arrive.) This particular curiosity was the pivot door. A pivot point was specified, usually about 12 in. to 24 in. in from one edge, and 4 ft. or so from the other. A 6-ft.-wide door was common, with heights up to 12 ft. We were weighing doors up to 350 lbs. The pivot hardware is easily found. The build was actually fairly easy, just large. The difficulty came in the weatherstrip and jamb design.

These doors recently gained popularity in California, in a sort of Mid-Century Modern Revival. The pivot door fit in wonderfully, and since it was California, with its mild climate, weatherstrip was secondary. The sill also had to have some height to meet the door bottom, to be 1 in. above the finish floor to allow room for a rug. I do not like to have silicone bulbs or similar running underfoot since they can snag and pull out and become a trip hazard. We elected to put two kerfs in the door edges, full perimeter. These could contact the jamb all along its length, twice, and make a good seal without much effort.

That is the balancing act—good, tight weatherseals and ease of operation. If we design the gap to be 1/8 in. between a door and a jamb, we can then size and use a bulb that will fit correctly and make a good seal. We have the silicone bulbs in four diameters, so if one is not a good fit, another bulb will be. There was no place for the Q-Lon or conventional compressible foam weatherstrip, so it all fell to the silicone bulbs.

Latching these pivot doors was as simple as a conventional latch. Or not. Since these were modern doors, and designs of a sort, some James Bond crept in, and some customers did not want the conventional, visible locks. One customer wanted to lock and unlock with her phone. Another wanted security, and no visible locks on the outside. We used magnetic locks, one low and one high, for a large pivot door that could be activated any number of ways, remotely or by keypad on the adjacent wall. Another door had a dead bolt with turn button on the inside, and no key cylinder or anything on the outside. I was never happy with the pivot doors, and now that it has been a year since we did one, I think I can say the fad has passed.

Secondary Doors

Secondary doors encompass screen doors, storm doors, and combination doors. The combination doors will have inserts of screen and glass for use in the appropriate seasonal situation. These doors are all secondary doors, used in conjunction with the primary doors.

First, a warning: A storm door with glass panels can act like a solar collector. Add a dark finished door, south facing, and temps can get to over 200° in the space between the two doors. This will age the finish, dry the wood, open joints, and eventually wear the door out. It does not protect the door—often the reason a storm door is employed. It is possible to get a UV blocker for the glass, and that is the preferred way I would build and install a storm door. The storm door manufacturers have for years recommended installing a wood storm door with a 1/2-in. gap at the top, and a 1-in. gap at the bottom to initiate a chimney effect that lets the heated air out the top. However, those gaps will reduce the energy savings one hopes to gain with such a secondary door.

When I first worked in a real woodshop, I was periodically given some deformed lumps of plastic that were originally moldings fastened to a metal door. Heat caused them literally to melt off the door. I was to make new molds by finding knives (amongst thousands of choices) to make the cuts, and then block out the parts, and do it so it would go through the shaper nicely. I was under the impression there had been a fire. After about the fourth or fifth set, I mentioned to the front desk that there seemed to be a lot of fires. They enjoyed the response, since it was merely heat buildup behind the storm door that had melted the plastic—no fire. Plenty of heat, probably over 280°, will do the job. From then on, I was smarter about what I was making, and I could excel at making the replacements.

Often, a secondary door is viewed by the homeowner as protection for the door we just delivered. Their first response is to cover it up, to protect it. I assure them that the finish will do its job, and the best thing an owner can do is look at the finish so they can keep up on the maintenance.

A secondary door is added often after the primary door, and is often hung on the exterior trim. The better installs will have the door swinging in a jamb, usually rabbeted. The most common mistake I have seen around doors is the mislocating of the latch. If care is not taken, the secondary door latch can interfere with the primary door latch. That is a hard one to repair. The damage to the installer's ego is usually more substantial than that to the door. The screen-only and storm-only doors are relatively easy to construct since they are rarely coped, and the square-edge design allows one to add wood panels. These doors are made from 5/4, and dress down to 1 in. thick. The glass is set in rabbeted edges with a bead of sealant on both sides to keep it all in place. Draw the sections first, and visualize the joints so you get the true picture of what you need to do. Glass for storm doors or panels is usually "double strength"—about 1/8 in., and add some for sealer. It must be tempered or laminated by law. If it is to be tempered, ask for the glass "no logo" to keep your work clean. Realize that laminated glass, at 1/4 in. thick, will weigh twice what tempered glass will.

Screen doors will require a wider and shallower rabbet (1/4 in. deep and 5/8 to 3/4 in. wide) than work that is made for glass. This allows for the staple fastening of the screen to the sash. We do not use the groove/kerf and roller method since it will not work on curves or with bronze wire. There are choices in screen material, with the most common being aluminum or fiberglass screen. We typically suggest and use bronze screen wire, as it is the most durable and attractive. It is not unusual to see places where the bronze screen wire has been in place for 50 years or more. In fact, we started using non-corrosive fasteners—stainless steel—when we realized the weak link was the tacks that rusted away in some old installations. The stainless staples will not react with the bronze, and will last generations if nothing else disturbs the installation.

A characteristic of our storm- and screen-only doors is that we hang them with lift-off hinges. These hinges allow you to simply open the door 90° and lift the hinge plate off the pintle or hinge pin. This allows for easy seasonal changes so a screen door can be stored in early fall, and then you do not have to deal with it all winter. Come spring, you drop it back onto its hinges and put it to use. Some folks want just a screen door, some want just a storm door. Many want both.

The combination door combines both a screen insert and a glass insert to provide year-round use of the secondary door. Invented in 1912 by what became the Combination Door Company, the basic door is a pair of stiles and a top and bottom rail. The inserts are sash, fit to the door frame by means of rabbets and hardware retainers. The inside edges of the door are rabbeted to the center of the door thickness by 3/8 in. wide. Matching rabbets in the sash are the same size. The sashes are set up to be installed from the inside of the building. The bottom rail rabbets are reversed on both the door and the sash, so as to catch the bottom of the sash and hold it in place while the hardware retainers are turned to secure the sash in place. It also works as a drain to prevent water from getting in the interior side of the door. Be sure to allow some clearance for paint or clear-coats. The retaining clips are available from the Phelps Company in New Hampshire. They are preferable since they are small, effective, and can be changed out with a coin.

Secondary doors of any type are out in the weather somewhat more than a primary door, so epoxy end-coat everything before and after assembly. Use brass or other non-corrosive type latches for longer life in the weather. Weatherstrip is not often used with secondary doors—

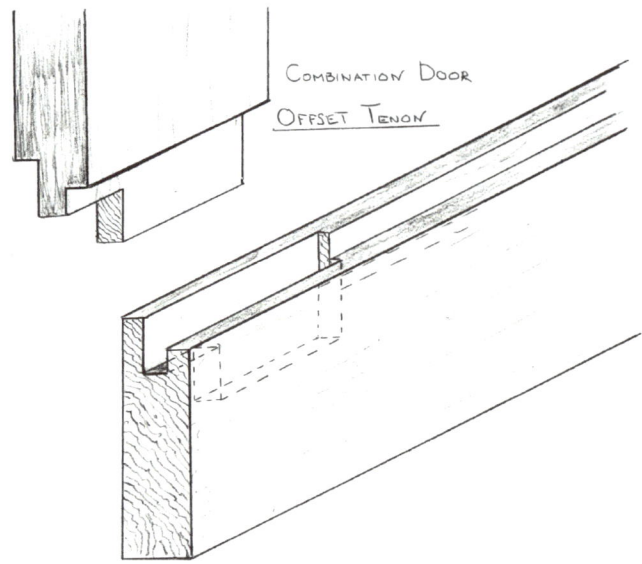

with one exception. We make screen doors for screened porches. These doors are to be well fit at all four sides. The jamb rabbets tend to three of the sides, with only the bottom rail without a rabbet. Mosquitos will find any gap at the door bottom, so we add a fin of neoprene rubber or a brush gasket in an aluminum channel, all fit into the plowed and epoxy-coated bottom rail of the door.

Secondary doors in residential use often call for a pneumatic door closer. They are traditional, expected. Seek and obtain a good one, since they can prevent a door from self destruction in a high wind. Set it according to the manufacturer, and adjust it properly. These closers are not as effective as the ones for larger doors, but a good one is satisfactory.

Good secondary door latches are a bit hard to find here in the 21st century. Deltana offers a good mortise-type latch that is all brass, and is non-corrosive in exterior use. Rocky Mountain offers a wider range of styles and finishes, and at seven times the cost of Deltana, they are the best of the lot. I am an avid supporter of craftsmanship and quality, but I struggle with the difference in cost, knowing the overseas craftspeople will never earn what we earn. The tubular latches for secondary doors, which are more readily seen, are now so poorly made that they do not function.

Dutch Doors

Dutch doors were fashionable in the late 1950s, with the Colonial Revival placing interesting things in the house the occupants had little use for. But Dutch doors do have a function. They are perfectly good exterior doors that can allow a horse to get its head into a kitchen and visit a while. Inside, they can keep young ones in a playroom, or a dog or two in a laundry room, while still allowing visibility.

In its simplest iteration, a Dutch door is a door cut in half, hinged with four or more hinges, and latched somewhat conventionally. If an exterior door, the two halves are rabbeted, then grooved for some sort of weatherseal. When cutting and rabbeting, think of how the door needs to operate—independent of the lower leaf. The upper section has to open and close onto the closed lower section. The lower can only be opened if the upper is open, or is opening with the lower. The door latch most often will go in the lower section, tying the two halves together, and then the door operates conventionally. A quadrant latch is the arcane hardware that is made for latching the two doors to each other, and it releases upon swinging the latch a bit. However, I see most Dutch door applications using a slide bolt of some sort. We have been asked to do interior Dutch door set-ups with a shelf. This is not a problem if the shelf overhangs on the knuckle side of the door, so it cannot swing into the jamb. If the shelf is more than a few inches wide, it will need to have brackets underneath to support the width.

Radius Plan Doors

Radius plan doors can be the most difficult door the door-maker builds in their career. They are a challenge, but not impossible, and are a great source of pride when completed. It all starts with a radius wall, and determining accurately the radius of that wall, and where the door will be in the wall.

How to determine the radius? When faced with a curved wall, work on the concave side if possible, and use an 8-ft. straight edge, with a tick mark in the center at 4 ft. Measure from the center of the straight stick, 90° to the surface of the wall (not the base mold, not shoe mold, etc.) and record that number, along with the length of the stick—96 in. Of course, the stick does not have to be 8 ft., but you do want a broad dimension to help the accuracy of your activities. Place your numbers into any of the many apps that will solve for the radius, check to be sure, and you are on your way.

One formula for calculating a radius is as follows:

a = altitude, b = one half the length of the chord, r = the radius

$$2 \times a \times r = a^2 + b^2$$

Insert the values for a and b and solve for r.

The radius doors you will be asked to build will mostly be 6 ft. or greater in radius. The door may be placed on the inside of the curved wall, making the inside of the door face flush with the wall surface adjacent, swinging towards the center of the curve. Draw a full-size section through your door, taking care to show the core radii and the skin radii. We would build radius rails and panels only. The stiles do not have enough radius to be too concerned with. I will usually leave them about 1/16 in. thicker, with 1/32 in. overhang to the rails, and fair the curve into the two faces once the doors are assembled. Be sure to think through the mortise/sticking placement on those thicker stiles. I could approximate that a radius wall of 8 ft. or more will not require thicker stiles, only a shorter radius will require stiles curved along their length.

The rails will be brick laid at the radius derived upon. Cleaned up, and with skins and sticking solids in place on the edges, the rails will look like they grew in place. The tenon end-work is done by propping up the off end of the rail at the tenoner to get the right angle of tenon. This angle is gotten from the full-size section drawing. Once the tenons are formed, do a dry fit of the door with no panels to see how the stiles sit in relation to the rails. Clamping should be strap clamp, taking care to ensure that the stiles do not rack or pinch, but pull up tightly on both inside and outside faces.

The sticking on the rails is made on the shaper if the curve is not too strong. They can be fed through on a cradle, similar to making stand-up curved molding. Using the same set-ups as you would with a flat door makes the radius door look just like the others, with the proper sticking, panel raises, etc. If the radius is too strong, then a router is used, with bits custom made to match the sticking on other doors. You may also find it advantageous to have panel-raised router bits made to fit the needs of the radius.

The radius plan panels are made from ripped and beveled stock, with staves much like a barrel. If the panels are to be stained, and they are a patterned wood like oak, then I suggest you start with a wide panel and rip the strips off of it at about 2 in. wide. Mark them as you go so you can preserve the grain sequence when it is glued back together. Set the joiner at the correct angle and make a light cut the full length of each edge, on each strip. Make two supports of 3/4-in. scrap that has the outer radius of the panel sawn into it. Place these two supports on the bench, securing them so that they are correctly in line with each other, then lay in your ripped strips. If you are happy with the angle ripped and jointed on the strips, then stay with it, spread glue and clamp, again with strap clamps. This type of work would usually be interior work, so I would use TBI or similar interior rated glue. When out of the clamps, sand the panel coarsely, working out the flats to fair the curve, and work up the grits to the preferred finish. Take your time. Make a 6-in.-long piece with a segment of the correct radius on it to slide along the face of the panels to see how the flats are fairing out. Sand as needed, checking frequently for true.

If this panel is to be seen, and you would like it to be further along the evolutionary scale, then the faces can be veneered, matched from panel to panel, and made even on both sides. A vacuum bag is the desired way to press such veneer. Remember to source this veneer from thick stock since, if the thin veneer is too thin, it is hard to press and will easily sand through.

In order to raise the panels on the shaper, make a curve cradle with the radii from both faces of the panels. Fix this cradle so your panel can slide over the cradle past the spindle loaded with the panel-raise cutters. A heavy router in a table will also work for panel-raising. Note that two cradles must be made, one for each panel face. One will have the cutter off the table, and the other will have the cutter low in the table. The sides of the panels will be raised a bit more conventionally, with the panel supported at an angle as it is shaped.

The same angle, in fact, that the rails used at the tenoner. Raising curved panels is the most challenging bit of millwork a craftsman can run into. Good tooling and good, secure setups will save the day.

For the two doors in the photos, I spent two complete days just going through the panel raises, making the jigs and fences, then processing them. Machining complete, we have mortise and tenon, sticking and copes, and raised panels. The door will still require some fiddling to get together dry. Be sure to have the two stiles rest on the bench, full-length, or you will be building a twist into the door. This is crucial—be sure the joints are all tight and that the edges of the stiles contact the bench. Since we know the bench is flat, we know the door is flat. The frame is easy if you have gotten this far: a head jamb sawn on the radii from solid stock with applied stop.

Radius doors are show stoppers in the shop or on the job site. I have built maybe 10 in my 50 years. Some were easier than others, they are not to be approached lightly.

Compound Curves

A curve within a curve is another way to say it. Either way, it can get complex. Visualize the inside of a silo, and the wall has a curve to it defined by its radius. Let's say an 8-ft. radius. Now we are gong to cut a hole for a window. The silo is 8 in. thick. When cutting the window hole, the worker got caught up and made an arch top window, with a full half-round top half. The jamb, casing, sash, and glass stop are all compound curves. The glass will have to be made on a compound curve also. Often the sash is left flat, the unit is set into a curved wall, and jamb extensions are made to bridge the gap from the flat to the curve. The casing is also a compound curve.

We will discuss the casing. The inside of the wall is a 96-in. radius. The wall is 4 in. thick, so that radius is 100 in. The window is 3 ft. wide, so the radius is 18 in. for the inside edge of the casing. We term the wall radius the horizontal radius, and the 18-in. radius for the window is a vertical radius. From this, you can visualize the two casing sets for this window. In order to make a part of the required radii, first make the horizontal radius part a sheet, bent on a form at 96 in. radius, and 44 in. wide and 3/4 in. thick. It is not necessary to make an entire "sheet" of wood for this, but it is helpful for visualizing. Next, draw the casing edges onto the sheet. This will locate where the radius must go. The verticals will not be too difficult to draw, but how can you draw the other radius?

Our solution is to draw the radius on kraft paper, and then cut the line, and lay it on your curved sheet, and you have the 18-in. radius drawn. This will give you a compound curved casing, ready to be profiled. And there I must stop. I have made simple S4S-type casings on compound curves, but I have no way to place a profile across the face of the casing blank. I can do straights, inside curves, and outside curves on the shaper, but not a compound curve. This we will leave for the CNC people. I would gladly let them make the casing should I ever need it for a compound curve.

Large Doors

Large doors is a relative term in this context, if there ever was one. Our daily bread might be 2¼ in. thick, 36 in. wide and 9 ft. tall. Usually a pair. So, when a door comes along that is quite a bit larger, then it is indeed a Large Door.

When we received an inquiry from an engineer who wanted a large and heavy door, 3½ in. thick by 54 in. by 116 in., he asked how we would build it. My first thought was to make hollow stiles, both as a weight and cost savings. But with all the joinery to run and assemble, I felt it would add cost back in, with not much of a gain, if any. We planned on solid rails and stiles and used two plies of 8/4 to get to the dimension. The panels were to be flat, 1¾ in. thick, retained by a 4⅜-in.-wide bolection mold. We opted to make the panel a stable one with thick veneer faces so we could match the grain. This was a 5-ply arrangement for balance.

The land for the bolection mold was built up to get it to the dimensions we needed. We let the two faces extend into plows in the stiles and rails. Plows that also aligned with the double mortise and tenon. I spent about a day making samples and mocking it up. But when the engineer saw the double tenons, he waved his hand and told us to proceed. It was obvious we knew far more about it than he ever could, and he wished us luck.

The point being that it is almost impossible to plan for a big door unless it is in your lap and begging attention. It is difficult to plan when it is unlikely that whatever does come your way will be exactly what you have pre-planned for. If you are given the option of getting to do the design work, then you can use the info here to get a jump on it.

With other large doors we have built, it often is a matter of finding enough panel shape and size variations to make the door(s) interesting. Check with the hardware vendors to insure the door can be latched. Rocky Mountain Hardware is one that offers custom fabrication, and they supply the longer screws, strike-plates, etc. that these doors will need. For the hinges on the door in the photos, we butted two 6x6 hinges tight next to each other, added finial tips, and got the Picard hinge look the designer was after. The larger Picard hinges we found could not carry the weight of this door. Speaking of which, we weighed most components, and estimate the weight of the door at 700 lbs. Once up and on hinges, it swung well and latched just as we'd hoped.

The old tale about the very heavy door that could be opened or closed with one finger is most likely myth. No matter what the weight of a door, the hinges are going to carry the weight, but they must do that with an uneven design load. The top hinge on that large door is under immense tension as the weight tries to pull the door down and away from the jamb. The bottom hinge has a lot of weight to carry, and it is under compression as the door weight levers into that hinge. I do not pretend to be an engineer, but I do not see what can or could be done to make a larger door that could be moved with a finger.

A year or two after the large door, I had occasion to go back to the house. I was eager to get a look at the Large Door, to see how is had fared in service. Well, it did open easily, and close and latch very easily. Surprisingly easily. Yet the great mass made itself apparent in that you could feel it as a massive door. Hitting the jamb and latching was a thud and a click, but in such a way that you knew it had mass. The owner was delighted, and even mentioned the "secret" that we knew to make large doors move easily.

Chapter Eight: Frames, Jambs, & Sills

Frames, Jambs, and Sills

We use "jamb" mostly for exterior type rabbeted jambs, and use "frame" for the interior type of jamb with an applied stop. But we do switch them around a bit, as you will see. Feel free to do the same.

Interior frames are typically 3/4 in. thick or thereabouts, with an applied stop nailed in place. The stops are not normally glued, giving the carpenter the option of "adjusting" the stops for a better fit. The frame itself is the width of the wall, plus about 1/16 in. Some shops bevel the frame edge this 1/16 in., while some leave it square. The additional width is to compensate for crooked walls or varying thickness of the wall. The heads are usually butted into a dado across the top of the side jamb and stapled, no glue.

We like our interior jambs square-edged, and about 1/16 in. wider than the wall. We are different because we miter the head and side together after the stops are attached. That is, locate and nail stops, then square-cut one end to over-length, then miter cut to length. Do it once or twice, and you'll get the hang of the best way to cut the wide miters. We also glue the joint. A couple of pins, and then a few countersunk screws crossing the joint will hold it securely. The pins should prevent creep along the miter faces.

Why miter? The butt joint incorporates a dado, and they are not consistent in depth from piece to piece. This makes for inconsistent jamb head lengths and ensuing clearance or door gaps. The miter controls this better, and makes for a better corner.

We have several masters in our work, and it pays to recall them periodically. The person writing the check is a master, probably the main guy. Another master is the building contractor who has a schedule, wants things right, and generally worries about the job down the road. Another master, less recognized, is the installing carpenter. If you want to hear a mournful sound, just send out pre-hung doors that are poorly fit. The carpenter will complain to anyone and everyone. And eventually it will get back to you, most likely too late to do anything about it.

So tune up your door hanging skills. Think about what you need to improve and what is adequate or better. Make a good door buck. Make new, tight routing jigs for the plates. Use the zero clearance method with flush top bearing cutters, and you can tune in the tightness before you ever make the first cut. For the routing of extension bolts, make a multipart jig that locates and then accepts adapters to make the deep plow, then the exit plow, and lastly the plate recess. Clean routes speak loudly of quality, so use the sharp bits, and learn to work carefully.

Interior doors, for the most part, now need to include enough clearance for air exchange. The inside jamb is cut 3/4 in. long—1/8 in. goes to the head for clearance, and that will leave a 5/8-in. clearance from door bottom to the door. Be sure to spell this out in the drawings or proposal—somewhere—or you may get caught in a pinch with someone who thought it should be 1/2 in.

The normal exterior jambs are made from 6/4—aiming for 1¼-in.-thick parts. We make these 1/16 in. to 1/8 in. wider than the wall thickness to accommodate for less than flat walls. The rabbet is made at 1/2 in. by 1¾ in. for a 1¾ in. sash. In actuality, we rabbet at almost 1¹³/₁₆ in. to give the sash a bit of recess from the edge of the frame. Doors will rabbet at 2⅛ in. to allow for the weatherstrip. The taller jambs—over 10 ft.—are now made from 8/4, and dress to 1¾ in., with the same dimensions given for the rabbets. While it is nice to have the extra 1/2 in. to work with, it tightens up the rough opening clearances since a half inch is lost there. We now make the brick mold 1/2 in. to 1 in. wider, and this cleans up the entire area.

Unfortunately, in this market, tradition has a mere stick of 2-in.-wide WM180 pattern brick mold for exterior trim. Nothing more. Sometimes an entry will have a nice limestone surround, and in that case, we use a 2-in.-wide cove molding and provide section drawings to the stone people. I have tried many times to get someone—anyone?—to brighten up and think about better trim scenarios than the stark minimum, but not much luck so far. Yes, they do say they are already spending too much on the entry, much less the trim. I think some of our more complex doors have suffered because the complexity stops at the jamb with common casings instead of a nice over-door with built up trim on either side, inside and out.

The foam weatherstrip requires a kerf and about ⅜ in. of room beyond the normal rabbet. The 1¾-in. rabbet that is really 1¹³/₁₆ in. now goes another 3/8 in. to 2³/₁₆ in. total. A 2¼-in. door rabbet will machine out to 2¹¹/₁₆ in. by 1/2 in. At one time, we tried to adjust the allowance for the compressible foam according to the size and weight of the doors. However, this quickly got very complicated, and we never felt we were accomplishing anything beyond busy work.

The compressible foam is a mediocre kit at best. Some doors fit well to it, and are relatively tight. Others barely kiss it, and the cold can flow in. We occasionally use a

"California" type weatherstrip that is a silicone bulb with a fin that fits right into the corner of the rabbet and easily crushes and seals as the door closes. The kerf can be sawn or routed, and we have four different diameters of silicone bulb so we can send out the best fit. This is new to our customers, and they are unsure if they should trust it, even though the mediocrity of the compressible foam, is widely recognized.

The exterior, or rabbeted, jamb is machined with two passes on a table saw, or a large straight cutter in a stand-up position on the shaper. Then the table saw is set for the weatherstrip kerf. Large diameter carbide-insert-type rabbet heads can be used with the jamb lying face down, as long as the exit corner is supported to prevent chipping out. The edge of the cut on the jamb can splinter if the tooling is not sharp. Give it a trial run, and see how it goes. You can use a dummy hardwood fence and bring the cutter through that fence to give 100% tight support at that corner.

Jambs are like stiles—they need to be straight in two dimensions. But the jambs get fastened to a wall and will receive the benefit of the wall making and keeping them straight. Nevertheless, your joiner skills will show with a pair of jambs. Well-made jambs help hanging go smoothly, and they ease the job of setting the door. Busting out and taking jambs to S4S is a good exercise in joiner skill development.

The addition of sidelights means that two jambs will be placed back to back or mulled to make a wider unit. For most makers, this means the crudeness of corrugated fasteners is used to fasten the two jambs to each other. The sills and heads are discontinuous, and the central door unit is not better for the attachments. The same is done for transoms. There has to be a better way.

A simple mull for us is to place two 1¼-in.-thick jambs back to back, glued and clamped after S4S, but still left long. This makes a 1½-to-2 ½-in.-thick assembly. We take care to glue and clean up the jambs so no "mull covers" are required. Those corrugated, nailed mulled jambs require mull covers to hide the ugly fasteners, and some builders still expect to see them until we point out the superiority of what we do. If so desired, by design or to fill space, the mulls can be made wider. This may be for the look, or to wrap framing.

When we want to have mulled jambs wider than the 1½-to-2½-in. minimum, we can do so with two strips (in each jamb assembly) milled to fit into shallow rabbets in both edges, or if it is more than an inch to be gained, then we will make a 1/4 in. by 1/4 in. plow and a tongue on the edge of a spacer. A 2-in. spacer will be 2½ in. with two gains of 1/4 in. each. The tongue and groove align everything so they can go together square and straight. Once the glue has set, scrape carefully, sand, cut to length, and join into the other members.

If the wall's rough framing is to be wrapped with the mulled jambs, we will glue the spacers into one side of the jamb, leaving the other side dry. A note will be made that this is to be a field joint. No need to fill the void with insulation, the framing inside the boxed mulls is fine. If there is a particularly massive door, you will need to be sure there is more wood where the hinges will go to take the longer screws and heavier load. Stout blocking should be glued and nailed into the areas needed. The void in the mulled assembly can be filled with rigid foam to prevent interior condensation. Solid or boxed mulls can be used in any vertical or horizontal placement that has sash or sash openings on both sides. Single jambs always frame the perimeter, and will tie the sidelights to the door jambs since it is one long piece, helping to hold everything flat, square, and rigid.

As the entries grow in size, it is prudent to make boxed jamb assemblies. Box beams are what they really are, and with glued joints, they are extremely rigid and stiff. Add the rigid foam insulation for the inside, and a box frame member will add stiffness and strength to any entry.

Joining two straight jambs is relatively easy. We use a simple "half of a half-lap" cut-out on the side jamb, and the head jamb is merely butted into that—the square-cut end of the head fits into the 1/2-in. recess made in the fat of the jamb. A bit of glue, and four to six cabinet screws, and the joint is made. Mulled jambs—solid or boxed—also fit to the head jamb with a similar half half-lap. In many cases, the head jamb is mulled to the bottom jamb of a transom. Any mulled verticals in the upper section will also fit to the head/bottom jamb with the half half-lap. This basic frame construction will serve in almost all situations you run across. Be sure to start and stay flat, and check square as you go. In the cut list, mark out the jambs that get weatherstrip at the door, and all the others will be the sash thickness plus 1/16 in. There is no weatherstrip at sidelights or transoms.

In recent years, we have seen the size of the entry elements increase; 42-in. and 48-in. doors are more common, and heights of 9 ft. and 10 ft. are seen regularly. Add larger sidelights—up to 42 in. recently—and transoms—two or three stages, reaching up another 10 ft., and it is possible to have a 10-ft.-wide and 12-ft.-high entry. Built in one piece it could not be moved, and it would never fit out the door of the shop. So we have learned exactly what will fit through the door, and how to make larger entries go together easily once on the job site.

Final assembly may happen on the front porch of the shop just before delivery in some cases. In others, we may elect to let the installing carpenters do the assembly. Our job becomes to design a method of site assembly that explains itself and cannot be built backwards or upside-down. Since most carpenters today have far more experience installing hung doors rather than hanging them, we have to define the line between assuming they know what they are doing and not giving them any latitude in how it is done.

We recommend white oak sills for most of our work. They will last nearly forever, finish well, and are the traditional premium sill seen in this area. Some of the older work will also have the heavy architectural bronze sills, now available from Pemko and Endura. The oak sill should be one piece, be about 1¾ in. thick, and have a 5° slope to the outside of the structure. It should be wider than the jambs by 1⅜ in. or so to catch the brick mold. This means the sill has "ears" that project beyond the width of the frame assembly. If the sill has to be glued for width, use epoxy on fresh-cut surfaces.

When calculating the frame size and drawing the detail to build, ask your customer what type of sill height they prefer. They will freeze up, since they have no idea, and have never thought about it before. We also are the only people who will ever offer them a choice. Some think ahead and want ADA thresholds in one or more entries. If those doors can be made to out-swing, then a rug can still be employed on the interior side of the door.

The typical homeowner will almost always have a walk-off rug at the entry, and it requires about 3/4 in. or more to avoid catching the door bottom. Add in the finished floor thickness, and the total sill height from the subfloor can be determined. The single-side jambs get half-lapped to meet the sill at 5°, and they are lag-screwed to the oak with glue. Mulled jambs are fit to the slope of the sill also, but cut and butted onto the sill. While a recess or shallow mortise for the jambs into the sill would be nice, it is a perfect place for water to work its way in and start the rot. Butting is fine. Screwed from below, and glued, these are now making a very rigid assembly. Mull covers, once used to cover those corrugated fasteners, are not needed since any mulled jambs are clean with no gaps. Mull covers now become a decorative item and can be half-round, fluted or reeded, carved, or columned.

The jamb drawing shows a sort of shorthand we developed to get the info we need into one place. The drawing describes a simple door frame with two sidelights. Overall width is given, as well as height from subfloor and finish floor to door clearance. Widths are also recorded. This frame has 2-in.-wide mulls and 3/4-in. outside jambs. The 36-in. door is a landmark, and the sidelights are given at 24 in. wide. This makes for a frame with outside dimensions of 89½ in. It is safe to say that this is a crucial dimension. It needs to be fed back to the builder so the opening can be properly made.

The height is similarly important. It has the added complexity of sill height location. A transom sash could easily be added to the drawing. Again, a walk off rug is requested, so we will accommodate one.

The intersection of sidelights with the sill requires an adequate amount of sealant in the bottom of the sash to seal that part of the unit. A 1/2 in. by 1 in. or wider piece of primary wood is molded on one edge, beveled to 5° on the other, cut tight to length and fit into the sill/sidelight connection along with sealant on all surfaces to seal this important place. Work neatly, but the inevitable squeeze-out can be removed after curing. The squeeze-out is your assurance that the seal is made.

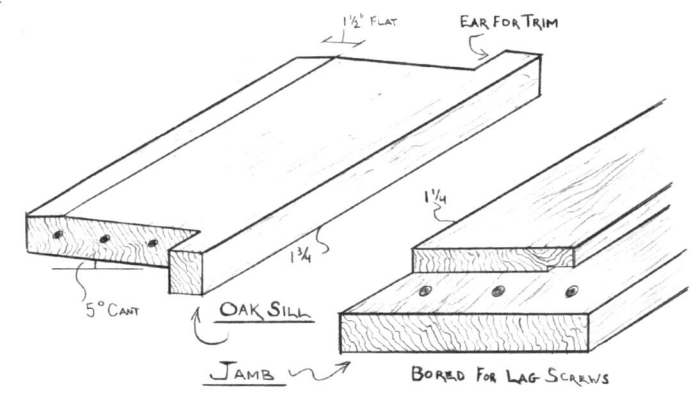

We get requests for entries that use aluminum sills on occasion. We stock an Endura 7-in.-wide, dark bronze sill with an oak adjustable riser for this work. Jambs are made up the same, but the ends that meet the sill are all cut the same, cut to fit onto the sill. A bit of RTV sealant is used to seal the end grain of the jambs, and they are screwed from below.

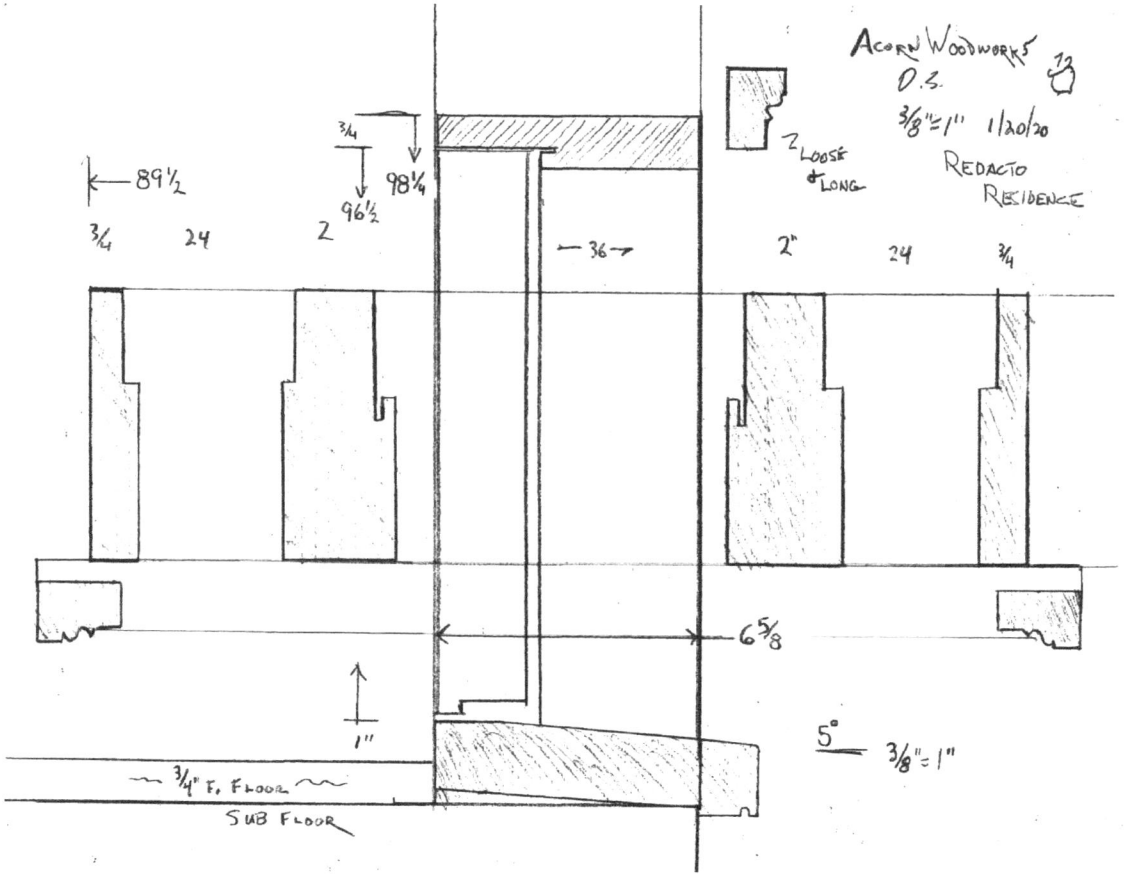

The oak riser is adjusted so it is located about midway in its range of adjustment at the door location. Sidelights get the oak riser removed, and a piece is made to hold sealant and fill the gap at the sidelight bottom while supporting the sidelight sash as needed. A beveled piece is made to deflect water at sidelights. Pairs require shifting the entire threshold forward about 3/4 in. so the extension bolt on the passive door can hit the center of the adjustable oak riser. We remove the factory-made flooring cover strip and replace it with one of adequate size to cover the gap made by scooting things forward.

Setting the sash into the openings is some close finicky work, but worth it for the look we get. The U.S. standard for sidelight and transom sashes sets them into deeper rabbets in the jambs, and then they are stopped in with 1/2 by 1/2 quarter-round or similar.

Acorn rabbets the frames 1/16 in. deeper than the sash thickness. That sets the plane of the sash and door just inside or below the frame. We run a 1/8 in. by 1/8 in. square "quirk bead" around the perimeter of every fixed sash—transoms and corner lights included. The recess creates a shadow that mimics the 1/8-in. clearance gap seen in the perimeter of the door and makes the interior sash stand out. The sashes are jointed and sawn to a snug fit into the frame. A bead of the silicone RTV is run in the corner of the frame rabbet, and the sash is nestled down into the frame and sealed in place. We will add a few screws to hold until the sealant takes a cure, and as assurance if the seal would ever fail and lose its grip on the sash. Sometimes a few clamps are used to help seat the sash into the sealant. And the squeeze-out on the other side can be removed after 24 hours. There is no way to clean up uncured silicone.

Sidelights and transoms are made with the same joinery and profiles as the door. They often align with the door elements, and indeed all the parts are made at the same time, with a good cut list. We have discussed most things needed to incorporate the added elements, so there are no surprises.

Take extra care in the drawing to insure you have everything aligned in the way it needs to be, and parts made at the same widths, etc. You need to be faithful to the architect's drawing, but they rarely have any detail beyond something just stuck in there. A horizontal and a vertical section are required, with detail drawings as needed. This drawing can be included in your shop drawings if the customer needs it, but usually these are only pertinent for the building of the project. My customers tend to only want to see elevations.

The frame should be examined in a design sense so you align vertical and horizontal elements in multi-opening frames and get the best look for your efforts. Usually, a door will be 36 in. or more, or paired, and height can be anything. Align glass sizes and openings horizontally and vertically so everything breaks on a line. Show it in scale on your drawings, uncluttered with numbers and symbols so the elements stand alone. Remember, your drawings are to inform the customer, the builder, the installing carpenter, and then the shop, so the project can be built.

A fine point to address for transoms is the horizontal head/sill jamb. The upper edge of the horizontal part is flat to the world and will hold water—never a good thing in woodwork. It is best to bevel off half the "fat" of the jamb. This leaves a good 1/2 in. for the transom sash to be sealed to, but then immediately slopes away from the sash to drain water away. Wider jambs will have a "flatter" slope, and narrow jambs will be a faster slope.

Water management is a running theme for all exterior wood work, and the frames we have been discussing are no exception. Use good woods, tight joinery, proper fasteners, and the right glues, and your work will survive well past decades.

This is a good place to discuss the "two sticks" method of measurement. You may be familiar with two sticks or you might have heard about it as a reliable way to gather data. One stick is horizontal. Plan on using all four sides. Tack it to the wall if possible, so you will have two hands free. Use a square, and transfer any points to your stick.

Label them as needed. Start with marking out where the major frame elements are that you can see—inside jamb width, vertical mulls, all full size. Give the stick a turn and you now have a fresh surface to indicate trim elements, sash and door stile widths, glass openings, etc.

The vertical stick is used the same way, marking out the major and minor elements. Once the sticks have recorded the info needed, then they can be taken back to the shop and drawings made, and then built. The sticks are kept until the job is built and then they can be used to check the work. Two sticks can be made for about any job, but the simple ones don't require the implied complexity of two sticks. Then, on the other hand, a large wall-panel job had over 140 L/F of horizontal stick and about 100 L/F of vertical. The job had about 2,400 sq. ft. of wall panels and was built with no mistakes.

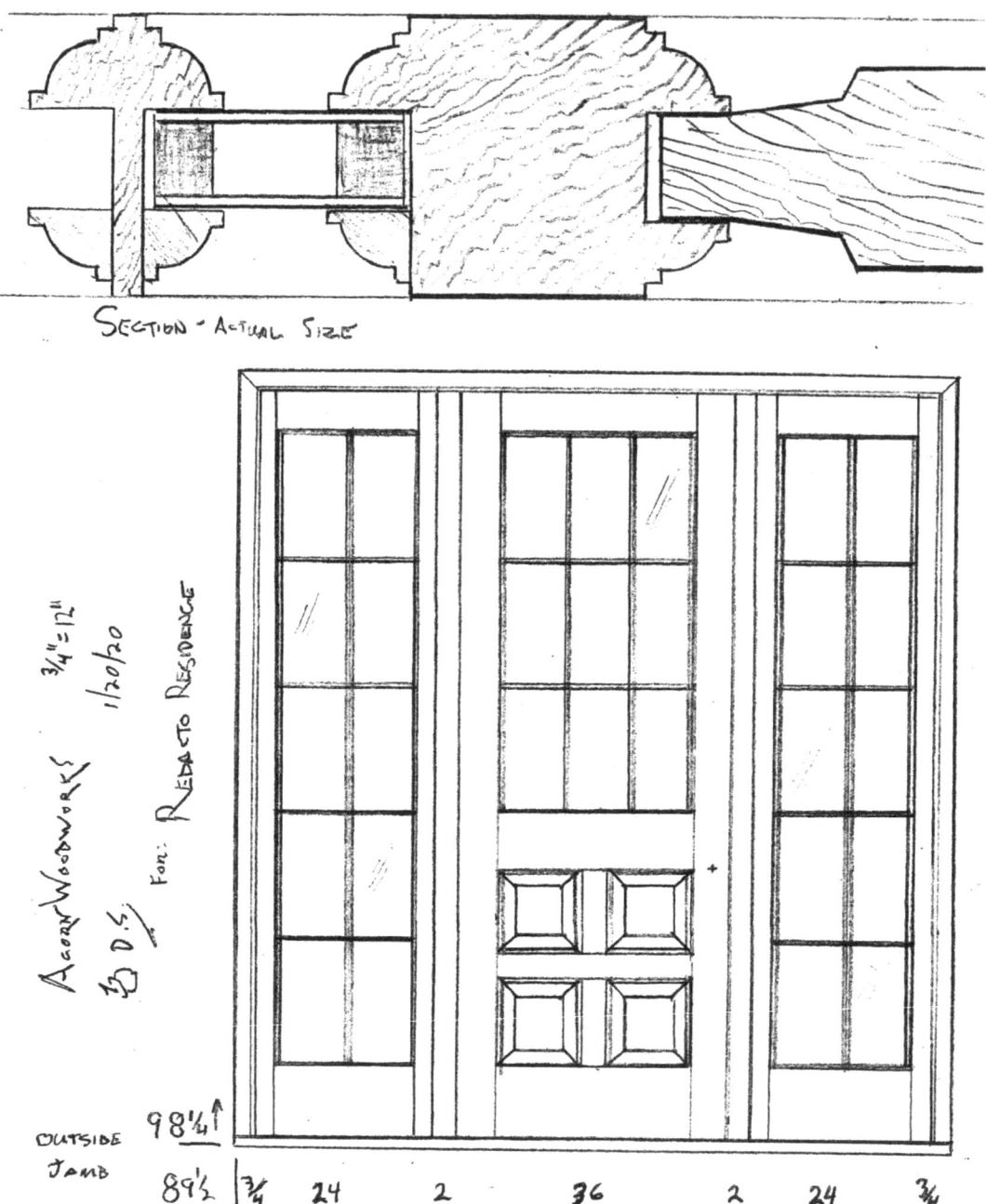

Chapter Nine: Curved Moldings, Rails, Muntins, & Frames

Curved Moldings

Doors are so rigidly rectangular that they almost beg for a way to break up the square and make things softer, easier on the eye. Curves were introduced not long after doors became more prevalent, perhaps in the 16th century. Since churches hosted the local wealth, they needed security to keep the gold and jewels inside the church. The doors that were made often had an arch added to the top since masonry accommodated arches better than flat heads. More accurately, two arches, culminating in a slight point. This design helped the masons since they could lay the stone in the arch with a simple form, and not a heavy header to transfer load. What we call a Gothic arch, or a two-curve arch, is part of the legacy that the designs of the 16th century have passed down to us centuries later. The heavy, single plank doors lent themselves to this innovation, and stylistically, the arch pointed up, a frequent characteristic of architectural elements in ecclesiastical work. Always leading the eye up, ever up.

Today, curves are in fashion for any number of reasons. Looking first at doors with curved openings for glass, we realize we need to make glass stops on a radius. All curves in this book will be defined by radius—the measurement from a center point to a point on the molding, usually the inside edge or the outside edge. A piece of 9/16-in. ovolo stop may have an inside radius of 14¾ in., therefore an outside radius of 15 in. A moment's thought tells us which way the molding curves. Simple

geometry is the only additional math you need to be familiar with. Nothing beyond first-semester geometry, and nothing to memorize. If you found demons in geometry class, you can relax. This geometry is fun.

I suggest that your first curves be drawn, even if they appear simple. Full-size drawings will help us with several things. First, they tell us how long a single piece of stock must be to get the needed part, as well as the width. That length is called the chord, and it will help us with the math and getting the wood out at the right size. Breaking the curve into two parts will make for shorter stock, and narrower, of course. But then, there is a joint to be dealt with in the molding. Joints can and should be avoided—they do take more time and can be seen as reducing the overall quality. Often, with long parts, or very short radii, it is difficult to get the stock in one piece, making two or more parts excusable.

The drawing can be checked against the work to ensure accuracy. The radius can be copied with a set of trammel sticks or large compass. Do not tape a pencil to the tape measure or use a piece of string. Do not ever use a bent batten for a curve that matters. The batten straightens out at the ends, so can not be trusted to be a fair curve.

Respect yourself, respect the work; make or buy the tools you need. A pair of trammel points can be purchased, and a stick made to hold them. However, one often has to have two to three radii to work with, and if you make your own trammel sticks, you can always combine them so you have any and all that you need to do the job. A set of shop-made telescoping trammel sticks will take a couple of hours, and last your entire career. With 3/4 in. by 3/4 in. and 3/4 in. x 1 in. square stock from a straight-grained board, yield out the following parts: two 12-in., two 36-in., two 60-in. and four 84-in. lengths. One of each length will be made into male dovetails (3/4 in. by 1 in.), and the other piece will be made into female dovetails. Work carefully. The dovetails can be made on the table saw, the router table, or even the shaper. Go for a good, sliding fit.

Once machined, sand them a bit, and then apply paste wax—two coats. Working with the female dovetailed parts, drill a small hole for a cabinet screw perhaps 2 in. long, fit so it is tight. It is beneficial to point the screw a bit on the grinder so it will stay in the center when needed. By using a removable screw, you can daisy-chain the sticks together and swing some big radii. Then, with the male sticks, go to the end and drill at 1/4 in. for a pencil, and band saw in from that end through the hole, and past it about an inch. Put a small screw in from one side of the cut through to the other, catching threads so the screw can tighten the hole and clamp the pencil in place.

We use a small spring clamp to hold the sticks in position. Hook your tape measure over the end of the bench, and hook the trammel stick screw over the edge also. Now you can easily line up the two and get an accurate setting for your stick.

All our glass stops are made on the shaper, with the necessary portion of the appropriate sticking tooling and a template. The template must be longer than the part to be shaped so you can ride the template—feed it first against the starting pin, then rotate into the cut until the template only—before the wood begins—is on the bearing. Once on the bearing, the template can be advanced, and then some curve parts.

Next, band saw the inside edge of the glass stock part, leaving the line, and cut to the center of the line on the template. Sand the template cut to remove the band saw marks and to fair the curve. Fasten the stop stock to the template and prepare to shape. DESTACO clamps are great for running multiple parts with the clamps fastened to the template. Brads, screws, and/or capture-blocking can also be used.

A hold-down from above will add a bit of pressure to push against. ensure that you can contact the spinning bearing with the template before you engage the wood and after. Feed into the rotation as usual. Once shaped, the stop can be band sawn to width with a single-point fence. Once sawn, it can be cut to length and dry fit into the door.

Regarding sawing curves on the band saw with a fence: Use a single point or finger fence. Those fences will have just a short area of contact, right at the blade, and will allow minor adjustments of the feed angle to get the width correct and stay on the fence. At some time you will be tempted to place a 7-in., 9-in., or 12-in. curved fence on the band saw and use that to get your width. While it appears logical, the fence will most likely fight you and not make a good part. The feed direction or angle has to be micro-adjusted as you feed, and the long curve will not allow that. It will be a fight, and your part will lose.

A word about band sawing curves. If you do not have a lot of experience with curves, you will be tempted to swing a router to create the templates or parts. Or use a CNC to describe an accurate curve. That is an acceptable way to work. But I suggest you learn to saw a fair curve. At first, you will feel awkward and incapable, but

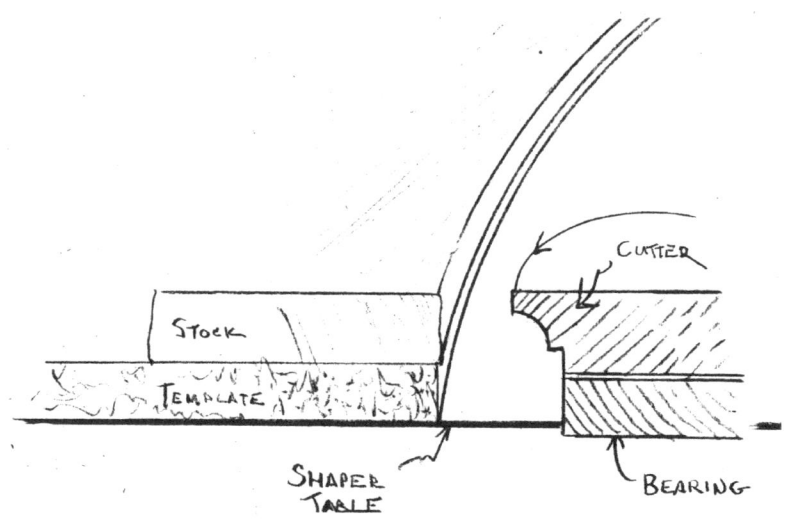

in time your ability will increase and your rejects will decrease. The goal is to look at a fine, curved pencil line and ask yourself if it needs to be sawn on the inside, the outside, or the center of that line, and then do it. In the meantime, keep some folded 60-grit with a 1/16-in. veneer slice in it to stiffen the sandpaper. Learn to feel with your fingers the high and low spots, and how to then remove those parts that are objectionable.

And there is double duty here: If the curve you sawed is bad enough, you can tip it into the bench vise and get out the spokeshave to fair up the curve and salvage the part. Improve on yet another skill. Sawing curves will come to you in time. Spokeshaving will also come to you. Swinging a router or using a CNC is all fine, but it is often not necessary. It rarely is the fastest, and we do have to remember that we are working with a limited time budget. Putting in the effort to train yourself is rewarded from then on every time you cut a curve. These skills you acquire will give you confidence and speed as you use them. They will never break down, abandon you, or eat G-code (the numerical programming the CNC router uses).

It easy to see how this method will work with even S-shapes and irregular shapes like ellipses, ovals, etc. We call these curves "flat" curves since they run while flat on the shaper table. Some moldings will not run flat on the shaper due to the complexity of the profile or the geometry of the molding. These we call "stand-up" moldings. Brick molding is a good example of "stand up" curved moldings. For these, we make a curved fence to match the outside radius of the molding, generally from stock sawn off in making the molding. This curved fence is attached to the shaper fence, and feather boards are added from above to keep the molding "down" on the shaper fence. The power feeder is brought in to help feed. It may only be able to track with two wheels, or even one, but with a good set-up, the feeder will do 100% of the feeding, freeing you up to feed parts and prevent calamity.

This is to be set up so the entire shaper fence and stand-up fence can be moved into the cut or out of the cut. Only the feather board(s) on the shaper table will need to be reset for the moves into the cutter. With the cutter out of the cut, you can dry feed the parts to ensure that they will feed properly without jamming. Since the cut can be minimized if so desired, it is good practice to run a part, or all the parts, at this reduced depth to lighten the load a full-depth cut makes (putting the part at risk), while checking that all points are functioning as needed. Once the cut is properly finessed, run all your parts.

The stand-up moldings have to run this way at a radius. Ellipses, ovals, S-curves, and free forms cannot be run stand up through a fence of fixed radii. This will take a completely different track. We will use a piece of scrap sheet—good ¾-in. is preferable; MDF is perfect. The blanks for our moldings will all be blanked out to thickness, width, and overlength, then fastened to the plywood from behind, with screws laid out to miss the portions of the molding that may be cut while making the molding. Router bits and a router are blocked up to sit level on the bench, with the moldings just underneath it. Complex moldings will require several router bits to make each portion of the cut in the moldings. A solid (no bearing required) finger is fastened to the underside of the router base to fit against the molding, positioning the cutter where required. This "bench routing" method is very workable as long as you have the bits required to make the cut. Most often, we will have custom bits made for the purpose.

Once your complex curves are made, they can be removed from the sheet goods and rabbeted, sanded, and cut to length for use in the door. Steam-bent moldings will distort from the water absorption and cooling, so cannot be used. Bent laminations can be used, but typically are considered over-working, since a good solid molding will work just fine if made right.

The curved moldings can now be mitered to the straights, but there is something else: "hunting miters." If the moldings are wide, or of a short radius, as they come together, the points of the shadow lines will not meet. Fine on the inside toe, and aligned at the outside heel, all should fall in place in the middle. But it does not. Now some moldings have so little difference, they can be sanded in and look fine. But wider stock and tighter radii make for the need to curve the miter. Scrap of both the straight and the curve are needed. Make the straight miters and set them aside. Draw the moldings carefully at full size, just as they are to intersect, and lightly draw the straight miter. Note that the shadow lines do not fall onto the miter line. Midway along the projected cut, the deflection will be the most. This creates a radius from the three points: the inside toe, outside heel, and this third point. These three make up a curve that can be used to modify the miters. This curve can be used to make a template to run the miters on a router, shaper, or even sander, cutting to the template. This is a lot of work, but if you do much curve work, you will need this one day, just as you are trying to make some molding come together, you will recall... The results are fantastic. Better yet, the work is correct. Not just "sanded" in. The above work is called a "hunting miter." Thanks to Gary Katz and his readers for giving a name to something I wrestled with more than once.

We may often be humbled by our work, but when you pull off a series of miters as described, you are at the top of your game and should behave accordingly: adult beverage, dance moves, rebel yell, a tear or two—all are appropriate. And now you know why we persist....

Curved Heads—Rails

When talking about "curved doors," one could be talking about a door with a curved rail that intersects a panel or glass—a curved opening, to be exact—or, one could be talking about a door with curved panels or glass, and is curved on the outer edge—so that the typical rectangular head jamb is replaced by one on radius. We will talk about radius rails and muntins, and then we will talk about radius head jambs.

Draw the proposed door at full size, accurately and completely all through the curved segment of the door. Add in the sticking, dotted lines for mortises, notes on the radius. In general, more info than you think you will need. It is imperative that all the curves in your door have the same origin or center point, so all the curves are concentric. This also simplifies the math, but more than anything, the rails are made with concentric curves, and they look right.

Give a thought or two to proportions. What are proportions you ask? Good question. Simply stated, proportions are values of size. The size of notepaper—8½x11—is considered a good proportion. Square is a good proportion. The French found that the Romans used a simple formula for their arch doors, and the Romans stole it from the Greeks, of course. That simple formula is: The width of the opening should equal the radius. Not the floor, not your cousin's forehead. If the opening is 6 ft. wide, then the radius of the frame should be 6 ft. also. It is what looks correct.

The full-size drawing will help you with your cut list so you can get the longer and wider rails that curved work requires. Most curved head doors fall into three categories: the half circle (quarter circle if a pair), segment arch, and either Gothic, elliptical, boxed, or two-radii arch.

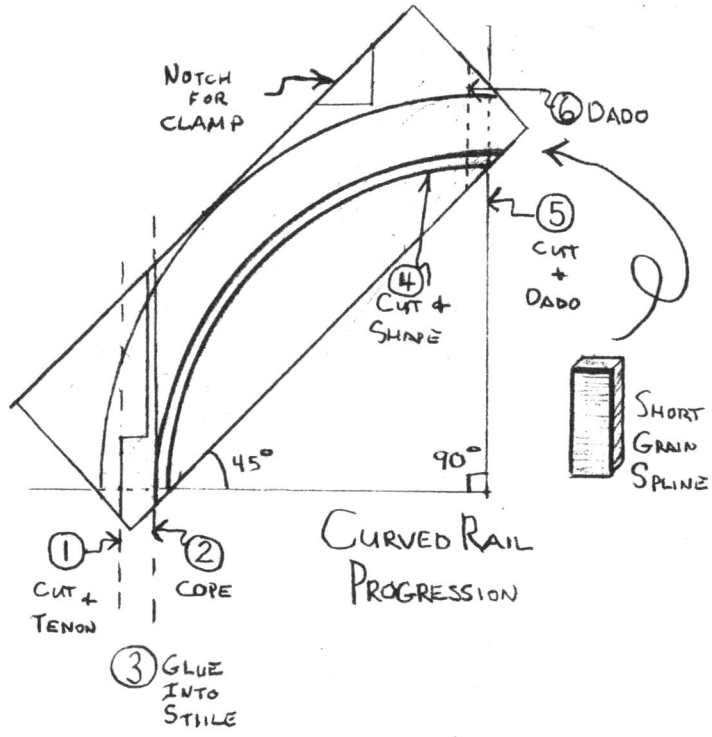

The half circle (for half circle head door) is drawn with the radius as half the width of the door. This lets the inside curve "finish" at the stiles where it all becomes straight. The circle curve cut into it is actually two rails, tenoned into the stiles at about 45°, and meeting at the center where they come together and get a full spline, glued with epoxy at assembly.

The curve is made of two rails, set at 45°, tenoned and coped into each stile. The curve is sawn, shaped, and rabbeted as needed. Then the center cut is made, fencing from the outside of the stile for accuracy. Once cut, a dado cut is made at what will be top center, and a short-grain spline is made to place at assembly.

This method helps give a better grain flow to the wood, and strength follows grain. It can easily be modified for Gothic or two-curve arches, and even flattened for elliptical heads. One occasionally sees arch-head doors with the top rails butted onto the end of the stiles. The rails may be vertical or angled, but are splined or doweled onto the stile ends. When these doors age and fail, the top arched rails wobble back and forth. They lose their ability to seal flat against weatherstripping, and then are destined for the trash pile. With the plan as drawn, the parts help reinforce each other, and the long joints keep everything stiff and aligned.

When the construction turns to an arched opening in a rectangular-head door, things get a little different. There is a problem when shaping the "start" of the curve and the "end" of it on the inside, where the curve diminishes into the sticking. This is fragile, and never can look good. It will likely come apart when cutting the cope since there is no wood to back up the cut. Parts flying about the spinning cope heads are not a desirable sight, so we will find a better way. Anticipate the problem with an attitude of resolution that finds a way to eliminate the problem. Lay out the point on your stiles where the rail/curve starts and stops. Cut into this point the width of the sticking (9/16 in.) and an additional 1/2 in. for a total of 1 1/16 in. Set up the shaper, joiner, or band saw to cut the 1 1/16 in. cleanly and squarely for the two stiles. The rails will have the 1 1/16 in. flat landing cut into them, and any curved layout will begin at this point. This point where the straight and curve change is called the springline.

A way to join the rail to the stile after cope, but before sticking, is shown in the drawing. Once the joint is cured, it can then be run as a curve on the shaper. This is an alternate to the butted sticking shown below, where curves and stiles are run separately. This is a fine joint if the cope and stick are good fits, since shaping after joining exposes the inner joint, and any insecurities it may hold. Like so many things in joining wood, there is more than one way to do most things. Either method is used for 90° intersections of curves and stiles, and neither would be used for other curves less than 90°.

Use your trammel points to draw the curves. The rails need to have sticking, overcut, and any other detail in the drawing so these can be incorporated into the whole easily; therefore, make them a bit wide, and lay out to include that extra width. Place the rail stock into the drawing, and draw the inside sticking radius directly onto the rail. Then swing the outside radius, if any. Establish the points at which the rail will intersect the stiles, and draw them in place. The point of drawing this all at full size is threefold: One, it helps you visualize the work as a whole; two, conflicts may be found and more easily resolved; and three, angles, part lengths, template sizes, and placements are easily found.

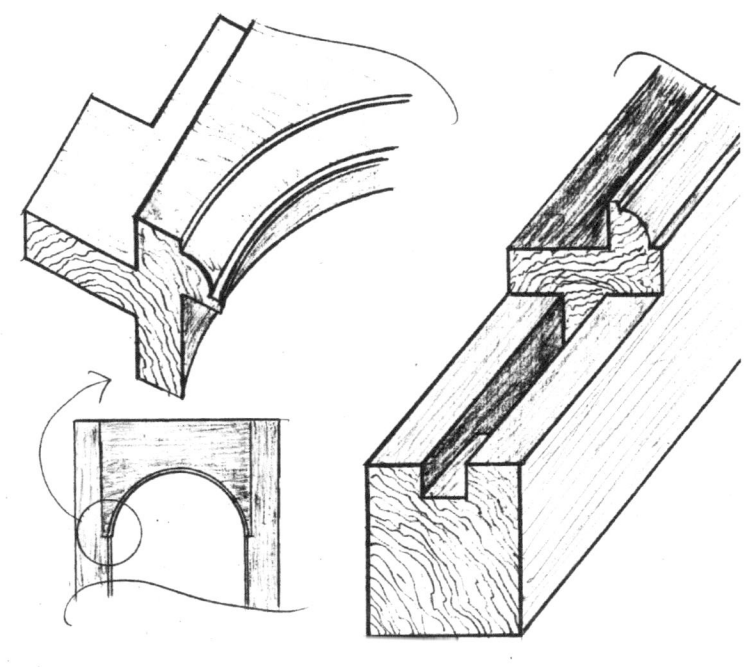

The two methods above can be interchanged as needed. Either one will suffice for 90% of the work we encounter. You will find one that works and may never try the other. However it is available, and there may be a time when it is just the ticket.

The tooling we designed has the same radius circle as the bearings. This means that, as long as we cut on the line or over, we will have wood where we need it. We will make a template from 1/4-in. to 3/4-in. MDF, either by hand on the band saw, or by swinging a router on a stick. If you do not have a lot of experience with curves, you will trust mostly the router and stick for its accuracy, but you will also know you need to get off the training wheels. As your skills improve, you may eschew the router and stick for the hand work that challenges us, then rewards us whenever we make a fair curve.

The preferred hand tools to fair a curve in the shop are a spokeshave and a stiffened piece of 60- or 80-grit. The sandpaper is folded over two or three times, with a thin strip of wood—1/16-in. veneer—in the fold to stiffen the paper and allow it to hit only the high points. The spokeshave needs to be very sharp and set to just hit the high parts. Shaving with the grain, template in the bench vise, taking thin shavings to fair the curve. Remember, anything on the template will show up on the rail, so take the time necessary to make a good template.

Once the template is made, the rail can be clamped, screwed, or nailed to the template and run on the shaper. Be sure the fastening is secure, and will not come loose under the pressure of the cut. Curved muntins are made the same way, but with a two-step template. The first will hold a part much wider—2 or 3 in. wider than the curved muntin. It will be fastened into the template firmly, with screws favored for this job. After the full profile is machined into the part, then locate it onto the second template that represents the other side of the part. Fastening here will entail more cleverness in order to hold it while a good percentage of its mass is cut away. DESTACO clamps, brads carefully placed, blocking securely fastened to the template, and receiving screws through the muntin spine are all ways to hold the muntin through the shaper.

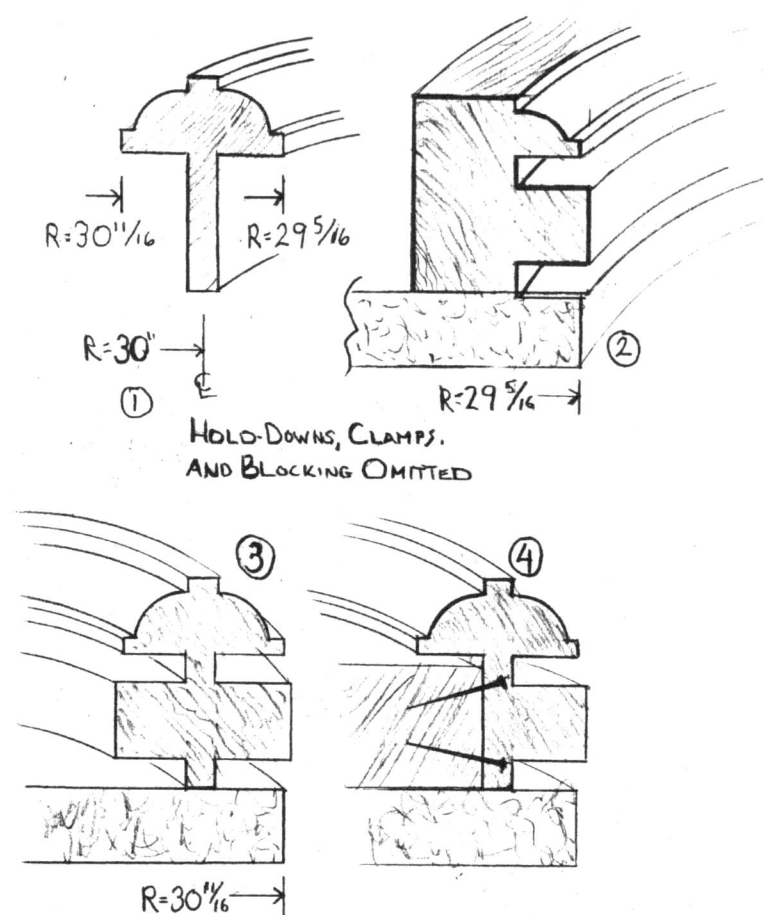

With curves, safety has to be ramped up and re-examined. A good way to look at the issue is to ask: "What happens to my hands if the part/template disappears?" A serious kickback will make the template and part disappear so quickly, you will never be able to react in time. Curved work implies fences moved, and more cutter exposed than with a straight cut, so an overhead hold-down may become a guard. A power feeder can be used—even passively—to help guard the cutters. I like to be able to see the cut—or at least some part of it, so I will use or make a guard that allows a visual confirmation of the cut. Fasten these things securely since they can become projectiles, just like the part you are shaping, were they to come loose.

Curved Muntins

Curved glass muntins look simple. But like so many simple things, there is much more going on. Just as with rails with curves on them, the end work of curved muntins is completed before the profiles are machined into place. The muntin plan is drawn onto the wood part clearly so the actual part is easy to use. The template is to be longer than the muntin, and cleaned up and fair to the curve. The illustration shows the steps in profiling, but it does not show the various ways the parts are held in place for machining.

The profile is too much wood to safely try to remove in one pass. So the cut is broken into two passes and two set-ups. Your tooling may be different, but the principle is the same—a lesser cut helps keep the parts in place and is less harrowing. By the time you get to the fourth set-up, you have a part that is considerably smaller than when you started. Fasten the parts securely, and take your time. Remember, this is why you make extra parts, and losing one or two is OK since you have back-ups.

If the tooling is one piece and cannot be broken up, you can make blocks that back up the radius in four, five, or six places along the part. Clamp down from above with screwed hold-downs. We have assumed tooling with a template and a bearing on the spindle. But there is another way. There always is another way in this work.

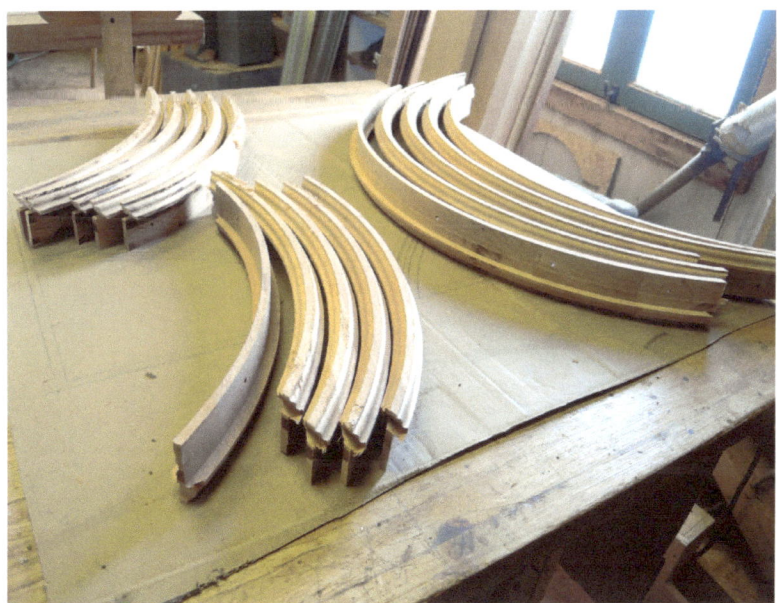

A *fixed* bearing is made of 3/4-in.-thick MDF with a smaller radius than needed by a few inches, and an oversize hole for the spindle, and runs well back to the back of the shaper for clamping there. It is to be set so it is flush with the cutters so it can clean the radius molding. It also can be set perhaps 1/2 in. further out—away from the cut. This will allow a minimal cut, lessening pressure on the part and the operator. Run the parts, move the fixed bearing back a bit, and run them again. Once you move the fixed bearing back to the flush point, you will have run the parts already, and you are making a minimal cut. The safe way to work.

Curved muntins typically intersect straight muntins for the grid-work required in the door. Make the copes on the straight muntins before they are profiled, and check the fit as you go. Again, a few extras will help remove the performance pressure. Build up the muntins on top of your drawing, aligning the real work with the drawn work to see where you are.

Once you have it all right and tight, then it is time to think about glass. I always looked at glass as the non-wood thing that has to be a part of the picture. But I now know it as a component, an ingredient that can add or subtract pressure. Draw the glass at 1/8 in. from the rabbet walls. You want that much space for the insulated glass. The two glass sheets can shift a bit, and the misalignment makes for a larger piece of glass than what you were planning.

You also want to hide the spacers from the insulated glass so this edge does not show. Asking your glass vendor for narrow, sight-line spacers is one step. Having your tooling made for 9/16-in. (or more) coverage is another step. The fact that we can hide the spacers and the insulating parts makes for a better insulated glass unit, and makes for a better door all around, and is something my competitors do not see as a problem.

The tooling in the photo is a 3/8-in. ogee cope and stick set so the glass will come in on the rabbets about 5/16 in. Make your templates carefully. If you have done your work well, then you may only need one or two templates. Openings that appear to be the same size, should be the same size. Order the glass from a competent vendor capable of fulfilling your requirements.

The sealant takes about 24 hours to cure, so the next day, the squeeze-out can be cut and cleaned off. This makes for an encapsulated glass unit, with the service life of the glass multiplied by two or more. More than one glass expert has said that our setting method will extend the life of the glass by at least doubling it. He was confident enough in our method that he did double the warranty of his insulated glass from five years to 15 years. I am unaware of an insulated unit failing in the 16 years we have been using the RTV silicone. With the previous product, a latex sealant, we would see a unit or two go bad every year.

Curved Frames

Curved exterior jambs or frames are built up from smaller parts and, in fact, are a great way to use up the short and narrow stock that tends to accumulate. These members are heavy in section and build, and indeed are structural elements in the completed unit.

Frames or jambs? Colloquial differences aside, the terms are interchangeable. One does not replace or augment the other. I understand "frames" better than I do "jambs," since I know what makes a frame. I don't know where jamb comes from, but the terms are used interchangeably in the shop. Yet another example of loose or regional language that is better used and understood than defined.

We make curved heads the same way for interior flat jambs and for exterior rabbeted jambs. We tried bent lamination, and found that trying to calculate spring-back, the number of plies, the thickness of each ply, etc., all became unwieldy. And once the head was taken from the form, it would spring back some amount, and no longer fit into our drawing. We have not tried steam bending for several reasons. Chiefly, I just can't allow fire and boiling water in an otherwise nice shop. The fire element is too much. Secondly, different woods, different ring orientations, will all bend differently, with different times for bending. Then we have to dry the frames out, and the resulting movement

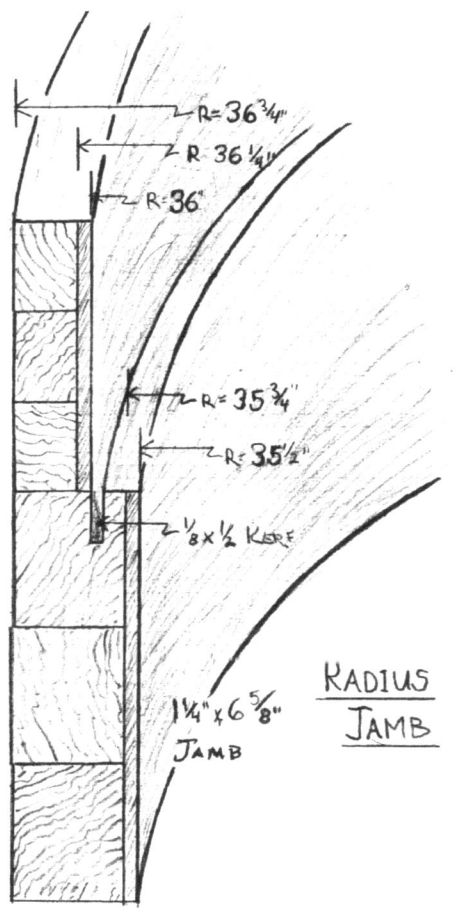

just cannot be counted upon to hold true to the radius or other parameters.

So, the problem became how to make a solid curved head that was dead on the radius, does not change radius, can be any width, any species, and will look right to even the discerning eye. The brick-laid curved head that developed as a result of that quest is the result we wanted. Brick-laid heads are now our standard.

Go to the job's drawing to determine the radii you will be working with, and the chord lengths. A chord is defined as a straight line between any two points of an arc. Radius and chord length are the two things we need to know for making most radius parts. For this discussion, we will use a radius of 36 in., for a pair of <36-in. doors. By "radius" I mean the rabbet into which the doors will fit. The inside radius will be 35½ in., with the door rabbet at 36 in., and the outside radius at 36¾ in. With two doors at 35¾ in., and a T-astragal at 1/2 in., we will have an inside chord length of 72 in. Plan on making a jamb that is easily 74 in.

The brick-laid heads we make use up the shorts and scraps instead of premium whole boards. Better to use the nicer cuts as the panels or feature rails. We will gather up a bunch of lumber shorts in the same species, thickness them all to the preferred thicknesses, then mark out a radius cut line on one of the pieces of wood—in this sample case, that line will be of a 35¾-in. radius. Move over 1½ in., and make another line on your part. Consider the values of "sameness" that will carry over to the templates that will be made to finesse these parts. The body of the head will be sawn at 35¾-in. radius, and the rabbet of the head will be 36 in., creating the rabbet. In reality, we can radius all the parts at the same radius—say 35¾ in.—and those parts that make the rabbet can easily bend to the slightly longer radius once it is made up and glued together. The thickness of the head parts is a low priority since, once it is all together, it can go through the planer to be brought into the proper width.

Cut out the parts. Sawing repeatedly at the band saw, try to stay on the line, but some wander is to be expected. Take a break and make up a 3/4-in. thick template for the body of the frame and for the inside of the head. Use DESTACO clamps, or screws, or nails, or whatever is needed to hold the parts in place while you feed them past a 6,600 rpm cutter head, shaping them into perfect little arches. I like to have a block at the tail to keep parts from backing up, and I will run two to three brads up through the template, so a 100-in. protrusion will hold things in place. The nails are directly under the DESTACO clamps so their collusion makes for additional security.

Load a straight cutter on the shaper, with a flush bearing immediately under the cutter head. Set a hold-down to lightly keep the part/jig on the table, and run a part. Repeat as needed. Since we sawed out the blanks at 1½ in. wide, roughly, we can set the inside edge over the template about 1/8 in. Clamp it, run it, and remove it. It should be a very fair curve, about 1⅜ in. wide. Run all your parts on the shaper so they are all clean and accurate. Get comfortable at the shaper, and run all the blocks.

You can elect to shape the outside radius of your parts if you like. There are two ways: The first is to use the shaper and make a template for the outside radius, then run all your parts so they are a consistent width. The second method would be to set up a point fence on the band saw, and saw them to a consistent width. It makes a neater finished project, but does not add to accuracy or longevity.

Once the parts are run, take them to the table saw or miter saw, and lay out the radial cut so the parts will butt together with little to no gap. Work through the parts, cutting each end. Build up your head on the bench, on a radial line clearly marked on the bench, kraft paper, or cardboard. Ensure that you have enough length, height, and layers to make what you need. The parts can be stacked with glue in-between and pinned in place. Continue until you are over-long a bit and over-width (height on the bench). Block up the assembly off the bench so you can get clamps under it and clamp up the assembly, checking for dead vertical as you go. Work slowly and carefully. We use TBII for this since it won't ever get wet, and it tacks up fairly quickly.

If you wish, you can make up the entire jamb at full width or over, and rip it into two parts to make the widths you need. This would be done before the thin liner is attached or after. The wider section that will be the meat of the jamb will need a "saw" kerf for weatherstrip on the edge that makes the short side of the rabbet. This 1/8-in. wide by 1/2-in. deep curved plow is made with a 1/8-in. spiral up cut router bit, and located just inside a 1/2-in. line, drawn as measured from the face. That drawn line is also the target for when the two curved components are assembled.

Once the assembly is dried, and the glue set, you can de-clamp and clean up the face, getting it ready for a 1/16-in. to 1/8-in.-thick face ply that will make the assembly look like one piece of wood. Too thick, and it may alter the radius it is built for. Too thin, and besides sand-through, it may telegraph the joints under it in time. Run the facings over-width by about 1/4 in. and over-long by a bit. Mark the center of the length of the face ply and center of the head assembly. Spread the glue with brush, roller, or trowel onto the assembly, and apply the face ply, starting at the center, working out to the ends so as to avoid making a "gather" in the process. Use a layer of corrugated cardboard against the face ply, then a layer of 1/4-in. MDF as cauls to press the face veneer down.

Once the glue has set for the liner, the two "halves" of the jamb can now be carefully glued together to make the jamb. It is now ready to be mitered to length. We miter curved heads to mate to miters on the vertical straights, as it is more accurate and reliable to miter than it is to make a series of rabbeted step joints. The geometry does not allow a butted joint.

We follow the same process for interior curved heads at 3/4 in. thick, both to build the head and to join it to other jamb parts. We get the same accurate and rigid results. These are a bit trickier to get right and keep right with the world as they are being assembled. Again, the results justify the effort.

If you do much of this work, I suggest you make an Angel-rig. It is named after an employee who developed it to help build up these brick-laid heads. It is a frame, approximately 3½ ft. by 7 ft., of 3 in. by 3/4 in. S4S on the perimeter and down the center, the long way, screwed

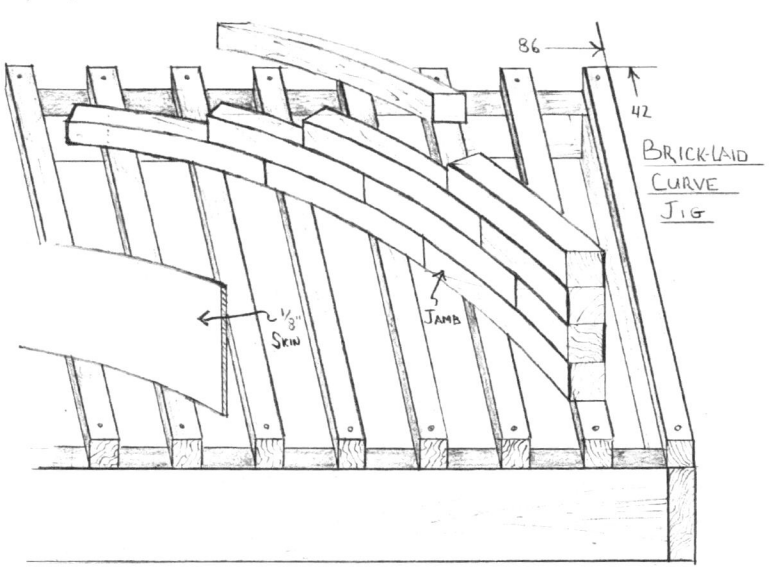

together at the corners, with 3/4-in. by 3/4-in. strips running the short direction, fastened about 1¼ in. apart. The surface of sorts created by the strips allows the radius to be swung with trammel sticks to guide the glue-up. The gaps allow slip clamps to fit in-between the strips and clamp the brick-laid parts in place. We typically forego the nailing since this works so well. The Angel-rig is a result of my long expounding upon creative problem-solving, and Steve really stepped it up with his solution.

The rabbet component of the head is usually just two layers of shaped parts, and on the same radius line as the body of the head. On occasion, the plan may call for an exterior rabbet on the frame. Usually these are for future or current storm or screen doors, with the better mounting on the jamb rather than on the brick mold.

Lastly, when joining a side jamb to the curved head, it is imperative that the two parts be co-planar, or in alignment. On a clean and flat bench, clamp the straight jamb to the bench, then bring in the curved head with as much of it as possible registering on the flat bench. Inspect the joint to see if it needs fitting to come up tight with the two elements in position. We do not want to deal with a "kink" in the jamb assembly. Once this joint proves out nicely, it can be predrilled, glued, and screwed. Start flat, stay flat.

Chapter Ten: Weatherstrip Hardware

Weatherstrip

Weatherstrip has changed greatly in the 50 years that I have observed it. I might say it has improved, but it is no great improvement. Once, we fit 1¾-in. doors to 1¾-in. rabbets, and then tacked in spring brass around three sides. There are many that still like—or it is "trust"—the spring brass. This leaf-type weatherstrip worked OK when installed by someone who could do so without kinks or buckles in the material. Unfortunately, no human being has ever been 100% successful at installing the stuff. It kinks with just one errant nail. And a standard door might take 200 nails! Even when not kinked, the metal does not conform to the door, and gaps are obvious. It conducts the cold and it is not unusual to see frost surrounding the door on the coldest days in the Midwest.

Today, the carpenter who wishes to use this weatherstrip has to buy and stock a van with all the widths of spring brass, plus the interlocks and V-molds, as well as brass nails and screws and anything else he may need. There is an expense to this. This is something the modern carpenter doesn't do: carry inventory. So we have taken this part of the work and added it to our scope, and we now supply whatever weatherstrip is needed. Supplying does not mean installing, however.

The current, or "modern" weatherstrip is a compressible foam that is ubiquitous in the trade. A vinyl jacket over a porous foam core with a plastic blade that fits into a

saw kerf, the compressible foam is called Q-Lon or Force 5. It is the standard, good or bad. We also send out "corner pads" for the bottom of the door rabbet, to seal at the juncture between vertical and horizontal. Our main objection to the compressible foam is the lack of a positive stop. That is, the door cannot be slammed. It just hits the foam quietly no matter how forcefully you close it. This "gushiness" is sometimes problematic with paired doors. We use concealed extension bolts for retaining the passive door in our pairs, and the top bolt may or may not hit the mating receiver bore for the door, depending upon the crushability of the foam. A 10-ft. by 42-in. by 2¼-in. oak door will crush much more than a 1¾-in. 3/0 by 8/0 knotty pine, four-panel door. So, do we make calculations for each door based upon size and weight? I don't think so. I prefer a better solution.

The soft weatherstrip also minimizes, or overrides, any effort we make at crowning a door. Crowning doors used to be common, and any carpenter would always hang with the crown. Simply put, the crown of the door is the side that is bowed up if placed on a flat surface. The purpose of a crowned door is a fine point of the hanging of a door. The belly, or fat part of the crown will go to the hinge knuckle side of the door, with the top and bottom of the door hitting the stops first as the door closes. As the door contacts the jamb at the top and bottom, the latch would be the last to contact. This puts a bit of tension on the latch, informing us by sound that it is closed. The tension will hold the bolt still, the door will not rattle, and it is tight against the rabbets.

Paired doors get a bit more confusing. We would crown the active door as usual, with the belly of the door to the inside. The passive door could have a reverse crown or be straight. A reverse crown is often thought to be too much tension. People do not like to have to slam or lean on a door to latch it.

Doors can be made with crowns if so desired. However, with the lack of solid stops to help make the crown effective, and the loss of knowledge to even work with a crown, one has to wonder why? We are pre-hanging the doors, so it is up to us to do it or not. Just as we do not ask what tenon length they prefer, there is no need to mention the crown.

A better solution for the future might include what we are now calling "California" weatherstrip as a nod to Gary Katz, noted carpenter and author, formerly based in California. This weatherstrip utilizes a silicone bulb—one of three to four diameters—with blades, available in 100-ft. rolls. A kerf is made in the corner of the two surfaces that create the rabbet, and a length of bulb is cut to fit. The blade is forced into the kerf and the weatherstrip is in. The door is closed on a business card and we see if the card can be withdrawn. If so, then a larger diameter bulb is placed. The weatherstrip needs at this point are draft-stopping, water-stopping, and light-proofing. We do not have to make an R-40 closure. One advantage of the silicone bulb is that the door can be slammed. That is, the door does go tight to the rabbeted stops and has a clean look. Indeed, it is invisible until the door is opened, much like the spring brass. But the solid stops are nice to have for crowning, and the door gives back a nice audible click when it closes tight.

The silicone bulbs, available from Resource Conservation Technology and Pemko, are not yet our standard until we can be sure that either "one size fits all" or whatever size we ship with a unit will work best. We do not want to ship out three different sizes, for instance. Resource Conservation Technology offers a router and holder that will cut the angled kerf and will even work into the corners of assembled frames. It is designed to work in the field. It is costly, but we bought one years ago for our explorations into different weatherstripping. A screen roller helps seat the stuff into the kerf. The fact that all this is different from what they are used to will alarm carpenters and cause them to balk when they have to set our work. We do support with instructional pages shipped with the hardware, but resistance is high.

Door bottoms require a more concentrated form of weather/water control. With the oak sills, we use a solid bronze threshold cap and brass J-hook on the door bottom. The door bottom is fit so it is a strong 1/8 in. from the oak sill on the interior side. Then a 3/8-in. deep rabbet is made on the rest of the door bottom.

The entire door bottom is painted with epoxy. Thinning up to 1:1 with acetone will make the epoxy even thinner and more apt to seal deeply. We will paint not only the door bottom, but also the sidelight bottoms and the transom bottom. This prevents water from traveling up the stiles (tree trunks not too long ago) and deteriorating the bottom rail/stile joints. Only the door gets the rabbet for the J-hook.

After the epoxy has cured, a length of J-hook is placed tight to the inside of the door bottom rabbet, cut to length and nailed in place, with nails about 1 in. apart. Watch out for kinking. It is best to place a length of wood scrap 1/8 in. by 2 in. wide by width of door into the J-hook to keep it from collapsing while handling the door. This will allow the door to be stood on its end just before it goes up onto the hinges. Remember to remove the strip of wood before closing the door.

The other half of this seal is the bronze threshold cap that is fastened onto the oak sill. Cut the cap to length, tight to the opening width. Close the door to latch. Place the cap into the opening at the door bottom, and push it back to ensure that it is snug into the J-hook all along its length. Mark the front edge of the sill cap with a pencil or point of a knife, clearly enough that you can get back to that point precisely. Draw a thin line of clear silicone along the bottom edges of the sill cap, on both the front and back "legs" of the cap. Place it carefully, and then drive the screws provided into the pre-drilled holes. The silicone is important as a barrier against driven rain. Omit it, and the door can leak. This threshold cap and hook also work as a door stop for the door, and will keep the door aligned and fit properly.

There may be situations where the cap is well seated into the hook, but the door still leaks with driven rain. For that, there is a 1/8-in.-diameter hollow silicone tube that can be set into the back of the J-hook. Draw a very fine bead of silicone onto the 1/8-in. tube, and carefully seat it into the J-hook. Let it cure, then reinstall. Then close the door to check fit. This should do the trick. If there is still a leak, then a rain diverter across the bottom of the door can be fit on. Next would be a porch roof, just after the lecture about exterior doors and how they should be sheltered since that is where people wait to go into or exit a building. Overhangs should be basic and required above any door.

Bronze sills are by far the finest, most durable sills available, as their expense demonstrates. They are offered in several widths and heights, all with the tongue for the J-hook, and several even have water return capabilities. A brass pan is placed under the sill, and drains direct any water that gets past the door bottom into the pan where it can drain to the outside of the building. We will make a white oak flat sill and fasten it to the jambs once it is coated with epoxy. The bronze is sized and fit in-between the jambs, set in silicone, and screwed to the oak with brass screws. Accurate weatherstripping has developed a modular system of bronze sill parts based upon a central riser that has water drains as well as the tongue that fits into the J-hook. Select the outer ramp, then the inner ramp, and the riser is selected by height. All the parts slide and fit together, and fasten with screws.

Paired doors on an aluminum sill get a vinyl bulb in an aluminum frame at the door bottoms. The bottoms are epoxy coated, as all door bottoms are, and the bottom gasket is sent loose and long, for field install. The oak riser in the aluminum sill is drilled and routed to receive the receiver plate and rod from the lower bolt of the passive door. We use concealed extension bolts on the edge of the passive door, and we use threaded rods to place the operational levers at about 24 in. off the floor. The upper one is about 5 ft. off the floor. These levers are much easier to reach than the less expensive "head and foot bolts" that our competitors fit to their doors, even when they are 8 ft. tall or more. We have the jigs and templates needed to do this work without hesitation. We will make all the routes and preps, then actually set the hardware—extension bolts and/or primary latch. Then we remove the hardware and box it back up. The doors are unfinished at this point, and the hardware should remain boxed until the doors are set, after finish.

Surface bolts can be used with paired doors, and we will offer to back it up with the extension bolts, or leave them off and just use upper and lower surface bolts. However, on 9-ft. doors, the upper bolt can run to 3 ft. long or longer. Not a good look in my opinion. We see surface bolts on fewer and fewer doors, just as the clavos/rustic fad is fading. Fortunately, we sold off all our speakeasy grilles just before they went out of fashion. I never did like when we were asked to put in clavos, and most times, I interjected my opinion that they need to be placed where they would have reinforced joints. While decorative, nails or clavos still have a functional reason to be in certain areas, while the decorators who know nothing about the history of nails in doors continue to put them wherever, with no logic at work.

A carpenter friend once had production in mind when he ran a trim crew. He would carry about 30 routers into a house, and each router was to be used as is, no changing of settings, no changing of bits—no wrenches in sight. This greatly aided production with no time consuming bit changes, set-ups, etc. Currently, we are assigning a task and a bit to a router for each and every task we do. No more changing bits for the bulk of door hanging.

Cremone bolts are a type of surface-mounted latching hardware once common with paired doors of the French style. The passive door has upper and lower rods actuated into keepers in the frame. The active door has a small latch that latches the door to the passive door. This hardware can be configured several ways. Most often, there is no hardware in the exterior of the doors, adding to security, and making them operable only from the interior of the building.

The bulk of our doors get concealed extension bolts, available from Deltana and Von Morris, among others.

The Von Morris is a better-made bit of hardware, but the Deltana costs about a third as much. We use extension bolts as opposed to "head and foot bolts" that fit at the top of the door and the bottom. Doors 8 ft. or taller would require a stepladder to open the upper latch. The bolts we use work with threaded rods, so we can place the levers at our ideals—24 in. off the floor and about 5 ft. off the floor.

We make a T-astragal for our paired exterior doors. The part is the exact same length as the hung door, and is nailed onto the passive door with brass nails. The astragal's function is to provide a stop for the active door. It has a kerf for the weatherstrip, and mortises for the locking plates for the extension bolts. The passive door is plowed—routed—on

the meeting stile about 5/8 in. wide and 1/2 in. deep for the extension rods, and recesses are routed for the exit plates for the extension bolts. The strike plate is also routed in at this time, at top and bottom, using a jig for accuracy.

Jigs are made and held to speed the work and develop accuracy. A plow has to be routed in the edge of the door, with pockets made for the levers, and end plates recessed at the top and bottom of the passive door.

We order most of our hinges from Deltana, a Chinese maker, but of good quality. Most hinges are now coming from China, and a good maker will keep up the quality. We specify square corner and architectural grade in our proposals, and often use ball bearings. We also ship dummy screws with the hinges for the inevitable take-the-door-down, put-the-door-up, take-it-down, etc. Once the door is up, to come down no more, then the nice brass screws can be inserted with no damage. Please pre-drill for any screws.

We fit and hang most of our exterior doors, and hingeing them was a real problem—and one I avoided if I could—until I resolved to improve the way we hung the doors. It turns out field carpenters do not want to hang doors—do not know how to fit doors, so that throws it back to us—the de facto experts. The first thing we developed was a horizontal hanging method. Instead of standing the doors and frame on end to fit and hang, we laid it all on the bench. Rectangular and single doors can be sized with a portable planer, then beveled on the strike side as the door thickness and hinge type require. Doors go onto their edges for planing and hinge routing, then are flipped for latch prep and beveling.

The second improvement we made was huge and truly gratifying. I gave away the Porter-Cable hinge routing jig that most carpenters learned to hang doors with. It was the reason they hated hanging doors. We evolved a hinge routing jig that used 1/2-in.-thick hinge-routing templates that were made the exact size of the open

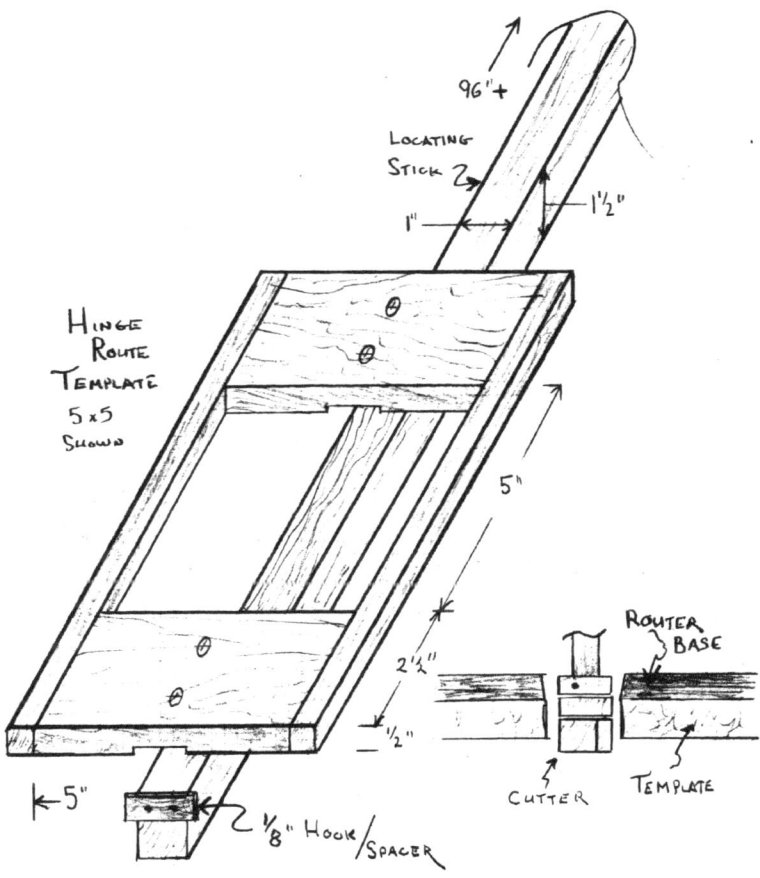

hinge. No more oversize metal templates, calculating, and hoping the bit worked well with the offset. The wood hinge route templates were made in 5-in., 4½-in., 4-in., and 3½-in. sizes, with about 10 pieces made for each size of hinge. The hinge template has a 1-in. wide, 1/8-in.-deep center dado running the length of the hinge template. This dado receives a 1-in.-wide long stick, and will hold all the hinge templates we want, where we want them, in alignment, spaced however we want. Nailed in place for a single door, screwed for more doors of the same layout, the templates are easy to place and easy to use.

A flush bearing top bit is used in a small router to route the hinges. It is a 1/4-in. shaft, with 1/2-in. cutting height, 5/8-in. diameter, and the bearing directly above the carbide cutters for a flush cut. This produces a clean cut without straining the routers, and it is very easy to set up. The hinge templates are placed along the stick, with the top of the stick butted to the inside top rabbet of the frame. This top end of the stick also has a 1/8-in.-thick offset used to hang over the top edge of the door for that route. This is a built-in 1/8-in. reveal at the top of the door. This simple method is so nice, I am surprised I have not seen it in use anywhere else. The hinges line up perfectly, and the hinge routes are easy to make. All measurements are direct, not indirect, and errors are eliminated.

Curved head frames require the door(s) be fit into the frame, and tick marks made onto the door and the frame to locate the first hinge. The templates can be located anywhere along the stick so there are infinite locations on the door. Since the stick is plowed into the underside of the template, it protrudes up a bit and helps back up the wood as the bit enters and exits the cut, preventing tear-out. The stick can be clamped to the door or jamb, and can even route jamb and door at the same time, with no loss of accuracy.

We build our doors a bit wide, or the frames a bit narrow. Close, but still needing to be fit. The doors are fit to the frame either by the joiner, or by the portable planer on the bench. The portable planer does the bevel. We only bevel the strike side of a door, never the hinge side. The setting carpenters may elect to bevel on the hinge side, but it is risky in that the geometry of the hinges is thrown off quite a bit, and it makes for a difficult-to-set latch. All of the fit and hang is done flat on the bench. Usually with some part of the frame overhanging the bench. The door is fit, the frame and door are hinge routed, hinges are temporarily mounted (using dummy screws), and the door is fit into the frame. Adjustments are made, if needed, and the door is then ready for latch prep.

While we use the typical knuckle hinges, we have had need at times for other types of hinges. Wide-throw hinges have been used to allow doors to fold back 180° to the wall. A 5x6 or 5x8 will throw the door out further on the hinges, allowing clearance. Most makers of wide-throw hinges will manufacture them with a thinner plate than the usual architectural grade, so the count on them may need to be increased. Some makers of wide-throw hinges call their hinges projection hinges. Work with a well-drawn section in order to determine what size hinge is needed. A 4x4 hinge will have about 1½ in. leaf to leaf at 180°. A 4x5 will have 2½ in., and 4x6 will have 3½ in. leaf to leaf. The 3½ in. will clear the door casing and allow 180° fold-back.

We also have had a call for Picard hinges. Picard hinges are similar to olive or bullet hinges, but they have barrels longer than the fastening plates or leaves. Most of this type of hinge are lift-off, and some include ball bearings. These hinges are not rated, but would be considered lightweight when compared to architectural grade butt hinges. In order to get the look of Picard hinges on heavy doors, we have taken 5-in. or even 6-in. hinges and stacked them tight to each other to get 10-in. or 12-in. hinges with steeple tips to add to the height.

If you are to use strap hinges, flag hinges, or similar primitive or wrought hinges, be sure you know how they should set, then preload them by moving the slack up or down, depending upon which leaf you are working with. These hinges, while perhaps suited to the décor and style of the client are not the best for architectural woodwork. They wear quickly, and will "sag" a door. The swinging action may also "spray" a line of dark metal where the butts rub. Caveats can be mentioned casually and clearly when the details are discussed, so if there are problems, they heard it first from the doormaker.

The "invisible" hinge is used for some modern work when the designer/owners did not want to see any hinges. The old and original SOSS hinges were the only options until the TECTUS hinges arrived about 10 years ago. Unlike the prone-to-sag SOSS, TECTUS hinges are heavy duty, sized for each particular door. and each hinge has three-way adjustability. It is the adjustability that makes the hinge attractive. Traditional hinges do not have any adjustability in them beyond a few cardboard shims. There is a video review of the hinges in use on Gary Katz's website: www.thisiscarpentry.com The hinge dealers all make the routing jigs and bits available so you can get a perfect fit with professional results each time.

The doors that have gotten TECTUS hinges, and the pivot doors that had minimalist latching/locking relied on hydraulic closers to hold the doors closed when unlocked. Clients recoiled when I suggested these, visualizing the ubiquitous metal boxes hanging off their beautiful doors. However, there is a broad selection of closers that can be fit into the floor, in the bottom of the door, or into the top of the door, or even into the head jamb. The closers are easy to adjust, and can close the door and hold it closed until it is locked. Most often these doors got a single-throw deadbolt with thumb-turn only on the inside.

We do a lot of latch prep because again, the carpenters do not like to do it. The complexity of the latches increases over time, and they have to relearn it every time they hang a door. An error in latch placement can be devastating for them. We have done several repairs over the years where we strip down a stile on both sides, full-width, and about 1/4 in. deep. We fill all holes and voids with blocks and epoxy and let it cure. Then we add the new skins both sides, admire our work, and then put in a latch in the right place. Cue Superman theme.

The latch must be in hand, in our shop, in order for us to install. Too many overly complex templates, multi-page directions, and bags of hardware to be easy to navigate. We have learned how to best handle the hardware, and now we can charge for it. When it goes to the job, it is easy for the carpenters to set. Some fiddling may be required, and is expected, but the bulk of the work, and the responsibility for it being correct, has been taken off their shoulders.

If the latch hardware is not selected/available when we build, then we make note and plan on putting it in after the door is set. We charge 2½ hours labor to do this in the shop, and double it for site work, plus travel time. Mortise locks are particularly difficult for the journeyman carpenters to handle since they do not have the Porter-Cable mortising router that makes quick work of a door latch mortise. Mortise latches can be set with routers and bits, but with quite a bit of difficulty and risk. Frequent use, as Acorn has, helps keep the skills up, times down, and the quality outstanding.

Tubular latches are best machined with a door boring kit. These kits will have several bits for the face bores and cross bores needed to set tubular latches. Routing templates are also a part of most of these kits. This is an important tool to have if you ever hang a door.

When we do a house full of interior doors, we will set up a line of operation stations, and move the doors along the line, hinge route, then bevel, then face bore, then cross bore, then face plate route and strike plate route. This equals efficiency, saves a measurable amount of time, and makes our work almost as fast as a pre-hung door machine, yet we can maintain better quality.

We are no longer dealers for latch hardware since we sell so few. I just ask the homeowner to provide the latch while giving them the name and number of our favorite hardware vendor. We locate the latches at the standard 36 in. off the floor, and mark drawings with the same. Some people think, since their new doors are so tall, that the latch should be set higher. I think they should only go higher once the doors are 12 ft. or over. Then the latch could move to 42 in. or even 48 in. off the floor. But, custom being custom, I'll put it anywhere, as long as they realize what they are asking for.

Multipoint latches are relatively new to the modern door scene. They send bolts or hooks up or down or out into receivers and hold the door closed at three or four points, a pair in five points. Their development and use came about in Europe as that region transitioned from wood door stiles to composition stiles. Composition components are not as stiff as natural wood. At the same time, European doors were increasing in height. This increase in height coupled with the change in materials made for a "floppy" stile that lacked the structural integrity of solid wood. A good wind could blow and bend the top of the door off the weatherstrip, making for a drafty, but tall door.

The first response was a surface bolt on the interior side of the door. This would hold the door closed, but made for cumbersome locking up and unlocking. The first multipoints offered only two shoot bolts that would exit the top and bottom of the door into the framework when the lever was actuated. Later iterations included rotating levers that swung into recessed pockets in the jamb rabbet. Then hooks came along.

Several drawbacks still affect the sale and use of the multipoints. Levers must be used as thumb-latches since knobs do not provide the leverage needed to actuate the linkage. Back-sets are narrow on almost all of the gear-latches/lock-body, with 2 in. being the widest commonly available, as compared to 2⅜ in., 2½ in., or 2¾ in., up to 3 in. for the conventional latches. And lastly, this hardware is changing all the time. Parts for latches that are just a few years old are no longer available. The hardware is still evolving, and rapid sunsetting is likely the norm for some time.

The multipoint latch is marketed as a security item since "floppy door" is not a selling point. Of course, the savvy buyer wonders why only a few imported doors will be so equipped, when many thousands of doors do not need the added "security." Security is usually an easy sell, and judging by the inquiries I get, a successful one. While we have fit it to their doors when a customer asks, we do not suggest it as a reasonable upgrade of any sort.

Chapter Eleven: Other Hinged Things

Gates

There are many things that are hinged, and may be doors, or may be door-like. Dog gates, baby gates, and yard fence gates are all hinged, and operate like doors. Overhead doors, driveway gates, and carriage house doors are also hinged, and operate sort of like doors, or in the case of overhead doors, look like they are conventional doors, but do not operate the same. These hinged things will be more frame than panel in most cases, but they are frame and panel, just like the doors we have been discussing.

But beyond all the different functions, all the above are built just like doors since they must carry their own weight, have a frame that gives strength to the assembly, and have a panel interior that is along for the ride. Driveway gates typically have hinges rated in excess of 500 lbs. to carry a gate that weighs under 500 lbs. Such a gate will have the top hinge in tension as the weight tries to pull it apart. And the bottom hinge is in compression, as the weight tries to smash it all together.

Even more in common is the fact that all these projects will require mortise and tenon joinery. It is the one joint that can be counted upon to hold under compression and tension, rack and twist. It is the joint from which the frame comes forth, and then the panel infill follows. Operating with a hand, or a remote control, the hinged panel is counted upon for strength, security, privacy, and more, just like the doors in the house.

Lightweight gates are simple enough. They are attractive and gets lots of favorable comments. The challenge usually comes in the top rail of the gate—curved, and/or pierced by pickets, and it needs to look right to satisfy the owner. Careful drawings are required. Then you have to follow the drawing, so don't draw something you will find difficult to make. A challenge is OK, but don't let it overwhelm you or the project. Other fence elements may be a part of the project, and they usually benefit from mortise and tenon construction.

For us, the most difficult type of gate or fence is the one with the rails pierced by pickets. Should the rails be split down the center, dadoed for the pickets, and reassembled so they can be continuous? Or should two-piece pickets terminate in mortises in each rail? One method assures perfect alignment; the other makes assembly easier. A curved rail, pierced, is difficult to make, but so is one with mortises both sides. Either way, it is work. In that respect, it is how we are paid.

When designing a gate, consider wind pressure and how it can overwhelm a person and slam the gate. Every large driveway gate we have made has electronic controls, so wind will not be able to destroy a slamming large gate. One pair had very few openings since the designer/owner did not like any design with openings. The wind did push them around even though they were on closers, and I have seen that they have been repaired twice. Draw in some openings for wind to pass, but don't lose your design. Just design with the wind in mind. Epoxy the end grain of panels, planks, and stiles. Build heavy enough so that when that chunky boy wants to go for a ride on the swing of the gate, the tenons won't fail.

There are many options for gate hinges, so be careful in selecting. Many options are decorative and are not in-tended to be operable. We often will use regular architectural butt hinges. They are all brass (no ball bearing hinges out in the weather) and will weather and age just fine. Strap hinges are popular for the look they lend. Some strap hinges are available as "fronts," and can be fastened tight to the butt hinges for looks. If you are using real strap hinges or loose pintles and hinges, pre-

load the hinge before fastening it, so the slop is already taken out of the fit, and the gate will not sag.

Gate latches come in all types from the primitive bolt and string to locking mortise sets. Gates tend to be thick—two to four feet—so be sure the latch you select will work. I am leery of prepping for a latch since I rarely know the hand of the gate, much less the exact dimensions that it will be set. A slight increase in clearance will render the latch inoperable.

As a result, when we are to supply the gate posts, we will often fix the distances so that the posts can only be set one way, and then the gate will fit perfectly. We will build the posts as hollow boxes about 3½ in. square—open—on the inside. The contractor is instructed to set two 2½-in. steel posts in concrete, x number of inches apart. The wood posts slip over the steel posts, and using spacers we make and supply, the posts are set permanently, usually with some construction adhesive and shims. Once cured, the posts are solid and ready for the gate and an operable latch.

While this amount of work seems excessive, we have found that it helps make our work function as well as intended, it eliminates installation problems, and saves money in the long run. This will also set us apart from the competition, which is often satisfied with their limited involvement. I feel that the smart business owner will seek and find ways to set their business apart from the others. With the lack of skilled tradesmen, does it also become our responsibility to insure our work is installed correctly? How extensive our post-fabrication involvement should be is a good topic for discussion. Or does our responsibility end at the loading gate?

Driveway gates are characterized by a wide reach, and limited height. This increases the tension trying to open the upper joints, and increases the compression on the lower joints. We will often double these joints, as we double the available glue surface area immediately. The members are larger—5x5 on up to 8x10. Dust off that chain saw mortiser and go to town. The cedar timbers you buy will be better quality and look good longer if you buy Free of Heart Center (FOHC). The timbers will crack less and be more attractive with no heart in them. Recently, we needed 200 L/F of 4x4 Western red cedar, kiln-dried. I could have ordered 13 or 14 16-ft.-long timbers, or I could order from my cut list and tell my vendor the 4x4 will cut into 18-in. lengths. This allowed him to give me an excess of my order, but in much shorter stock that they had lying around, and frankly were glad to get rid of. Be your vendor's friend and work with them. As they learn your needs, they should be able to better serve your company.

Driveway gates are often an open pallet for a design of fretwork or metalwork. This means working closely with the metal craftsman to integrate his/her work with your own. Find a way to integrate the metal so it will not create places for water to trap or cause problems when it moves in an opposite way than the wood.

Shutters

Shutters are as popular as ever, and a good item for every woodshop to produce. Around here, we see the three boards/two batten type shutter screwed to walls everywhere. The sizing is less than casual, and if there is a radius, it also is often wrong. It is just a terrible state of craft, but no one seems to mind. I have had conversations with owners who approve such drivel because it is the end of the project, they are out of nickels and dimes by now, and they just want to be done.

However, if you can get approvals earlier, you may be able to avoid that conversation. Besides the board and batten shutters, we also see frame and flat panel, frame and raised panel, and louvers. No matter the style, shutters should be mounted on shutter hinges, and retained by bullet catches or tiebacks. Mounting the shutters on hinges makes painting and maintenance much easier, and it allows air and water to pass behind the shutters, therefore prolonging life. If you have a good source for the hardware, then you can quote it easily, and make your pitch for better shutters. We use Timberlane Hardware for the shutter hardware. You can find many styles and types there, as well as some styles in black-painted stainless steel.

Board and batten can have the battens on the "face"— the side normally exposed when the shutters are open. Or they can be concealed on the back. Glue should be used, as well as stainless screws to back up the lack of joinery. Plug the screws for a clean look, and don't forget to give your boards a bit of clearance. Less if they are V-groove tongue and groove, more if there is no tongue and groove, just boards.

Frame and panel are simple enough. If there is to be a mid-rail, it should be aligned with the mid-rail of a double-hung window. In the 70s, I learned that the windows with a higher mid-rail are called "oriel," and good practice stated the mid-rail needed to match that rail. But now I am told "oriel" means a bay window, with nothing to do with mid-rail location. So much for regional language. No matter, the rails still need to align for the best work. The brick-to-brick width can be divided by two for the net functional width of the shutter panels, and from the brick sill up to the header, just inside is the functional height. A proper touch for the meeting rails is a bead and rabbet, about 1/2 in. wide, just as seen in functional shutters to blind the center when closed. This complicates the width measurement slightly.

With flat panel shutters, it may be possible to use a man made panel and glue it in place. Extira—the waterproof MDF may be a good choice. Realize it is water resistant, not waterproof, and so it needs to be painted well. Exterior plywood of an exterior grade can also be used, but is rarely seen today other than 3/4-in. fir tongue and groove—too rough for fine millwork. Medium Density Overlay (MDO) is a no-void fir plywood with resorcinol glue, often used for signs. Both sides are covered with a heavy kraft paper giving a smooth no-sand type face.

Raised panels are made the same as basic frame and panel, with epoxy end grain on the panels. Be sure to get a proper joint on the panels, and then proper glue for width as needed. I like to keep the panels at ¾ in. thick or thicker, so it may be necessary to run both sides of the panel to get it to fit correctly.

Louvers are still popular here in the Midwest, though fewer and fewer people seem to recognize them. They have more popularity in the South, perhaps because the Northern climes did not need louvers so much as they needed more solid protection over the windows. Louvers are measured for and built the same as other shutters. Note that when they are closed on a window, the slats are down and in a good position to shed water. That means the slats are "up" when the louver panel is open. This is contrary to what you will see, since the original reason for shutters has been lost. False and real moveable shutters will have a pushrod up the center on the louvers to actuate the louvers when the panels are closed. Use corrosion-proof fasteners and pins when making real or false movable shutters.

Overhead Doors

Overhead doors are best if made into horizontal panels like the older wooden overhead doors of the 40s through the 80s. The problem is, designers today like the look of carriage house doors that were side-hinged like a passage door, and had more of a vertical design than horizontal. These can be a challenge to pull off. Since these types of doors are often one-sided, with the one face of the door as public and exposed, and the other side private and only semi-exposed. You might think of building a series of overhead, horizontal panels, with false fronts that show the desired look. The backs can be filled and blocked as needed.

The popular overhead door today is a commercial door of some modest type, with some tongue and groove "panels" and 2x "frame" on the perimeter. This is all glued on with construction adhesives and sometimes pinned. Then about three to four years later, the boards start to come loose, so the carpenter is called out to glue and affix the fallen wood. These doors will fit within a "budget" of sorts, but the added expense for the future is not calculated.

Historically, as garages grew in prominence in the postwar boom of the 1950s, woodworkers merely made a frame-and-panel assembly about 24 in. wide/tall and 8 ft. long, hinged four of them together, and there was a good, serviceable overhead door. This long, horizontal look was well suited for the ranch design prevalent in the 50s, and so heavily influenced by the Prairie School and the emphasis on the horizontal. Today, the designs are more eclectic, more "European," and more demanding. The four-or-five-panel horizontal is not the baseline anymore.

With doors that now have a two-door hinged look—the carriage house type—but still function as overhead, we need to make a frame that is stable within each of the four or five panels that will hinge together. At its most basic, a joined frame can be made approximately 8 ft. by 8 ft. with intermediate rails running horizontally, where the long hinge joints will be. Tongue and groove boards can then be applied, glued, clamped, screwed, and plugged from the back. Once the entire panel is completed, it can be sawn apart horizontally at a 10° bevel. The bevel is to prevent seeing into the joint, and the angle is to drain water. This is also a good place for a tapered tongue in a groove joint. The tongue and groove will keep the panels aligned and snug all along their length, and water will be prevented from penetrating the door. Don't forget, the tongue should be on the upper edge, and the groove on the lower edge.

A second method involves making blind stable frames that are concealed enough to allow a frame-and-panel design that can support a façade that satisfies the look required, but with a design that will allow the wood to be wood. This may entail veneered plywood for stability, with frame materials either surrounding or hidden behind the panel. Angled tongue and groove joints will be imperative as proof of design skill and commitment to long-term exposure. Draw these details out so you can resolve them on paper. It is much more rewarding to go to the shop with everything known than it is to have a number of things to figure out before building.

Chapter Twelve: Problem Solving

Woven Wood Panels

The job was pair of 36-in.-by-108-in., one-panel Honduras mahogany doors with a "woven look" to the panels. I think I may have seen this woven look once before, perhaps in an Indonesian gate. I think that gate was actually woven, using green wood. The request came from an interior designer who knew the shop and would let me fill in the details, as long as she was kept in the loop. The only other particulars were size. The doors were to be the main entry for a new upscale restaurant.

The problem solving—one of our favorite things—started right away. I found myself sketching at the dinner table, at the brewhouse, at my bench, while I should have been doing something else. The complex projects get into my head and take over. Not at all unpleasant, but hard to avoid. First up was some sketching, and I spent quite some time trying to draw the look, and to visualize what we were to make. I had to close my eyes and rotate the parts in my head, and hold them there, like in a meditative focus. Then I'd get distracted, and have to bring up the images again, and hold.

Eventually, I knew enough to draw again. The drawings were nice, but only captured two dimensions—they lacked that pesky third dimension, critical to a woven look. So, I moved on to wood, settling on 6-in. squares of 8/4 poplar and maple to shape and play with, as well as some MDF squares for the two-dimensional effect. I then started thinking in terms of tiles—a major step towards a solution. Prior to that, I felt the parts were to be one-sided and reversible. One part would be an "out" part, and the neighboring part would be an "in" part. Turn them over, and the ins become the outs, and the outs turned into ins. That presents some interesting characteristics, and it may be worth pursuing one day. But not this job, not this time.

Now the next step was Justin's realization that squares were not the best way there, so we changed it to where the tiles were 3/4 in. longer than wide, or 3/4 in. wider than long. First with some MDF mock-ups, then, once we saw how that helped things, we moved on to some 8/4 poplar tiles. The tiles would all have grain running vertically in the door, although this is not the only way to go. We did it so we could treat the panel as a solid wood panel, four tiles wide and 12 tiles tall, all grain running vertically.

I also made the width 3/4 in. narrower than the length. This made for a small 3/4-in. opening as the parts—now tiles—were arranged. We had a fractional bit of latitude in panel width, but had to make the parts come out even as they ran from stile to stile. We had more latitude with the height by varying the bottom rail width. We could settle on a master part size fairly easily and early in the build.

I quickly sized about six to eight parts and began to manipulate them to see how they might work. I added a 45° bevel to the four long edges to give the illusion of a taper as the parts weaved into and out of each other. The bevel could be modified should it need it.

I thought I was making great headway, but the limitations—two dimensions, with only a hint of the third—presented results that did not look like a woven panel. The sides were to have plows, and two would have tongues. As I started roughing that in, I realized something critical: These were looking more and more like tiles. Specifically, I found myself thinking of the old 1/2-in. thick ubiquitous parquet flooring that was made of parts of smaller wood, but they all had two tongue sides and two groove sides adjacent, so they could be pivoted around just like tiles. That seemed to work. It worked so

well in fact, that I was sure I was going to have to discard it since the woven part would probably preclude the tile solution. The third dimension.

In our normal design work, we have only two dimensions. The third dimension is implied and fixed in the door thickness. It is almost never a variable in a design sense. So we have these nice 2¼-in.-thick door frames, and in order to enhance the depth, I planned on using 10/4 mahogany for the basket weave. I was forcing myself to work with the scrap tiles since the Honduras was a bit expensive to play around with.

As I was visualizing one side bowed, I sawed a curve on one face. Then on the other side. Then the same on a second tile. Serendipity arrived, with a quick, overwhelming and warming realization that I had it—no need for a concave side for every convex side. They can all be convex on both sides of the part! I did not need a bowed side and a hollow side, only bowed. The concave is implied, but never visible. Never made.

That rush of awareness, that stimulation of the heart and brain is at the heart of why I work wood. It does not happen every day, and is, in fact, pretty damn rare. But the several times a year that I get such a feeling are wonderful and keep me coming back.

I think if we are to do it again, we might alternate the grain for an even more accurate woven look and feel.

We beveled the edges on the poplar, giving a dimension to the "sides" of the weaving strips. This eliminated the integral tongues, a step I was reluctant to give up since I appreciated its elegance and the need to make bunches of loose splines. The faces of the tiles were curved to a 10-in. radius segment. This gave us an apparent change in thickness of 3/4-in. per side. How to make the curves? This had been discussed quite a bit during breaks and general conversation.

We began to build a 20-in. circular jig that would hold four of the tiles at a time and rotate against a fixed bearing under some 6-in. straight knives in the shaper. The 20-in. diameter of the jig coincided with the 10-in. radius that gave us a radius of for the face of each part, or tile. A full plywood top for the jig further helped hold the tiles in place while the power feeder fed them in a slow, accurate circle. The tiles were held in place on all four edges. At the time, we felt we had this dog licked, that the jig would do everything but fix lunch.

The Jig of Complete Happiness was immediately a problem. The cut was not huge, but soon proved too much.

The Jig of Doom as it came to be called once we found its limitations, while very well executed and stout, had a fatal flaw. Cutting into the grain was just too much for the tile, and we saw a 50% failure rate on the first go-round of four parts.

A very dramatic failure rate as evidenced by the sounds and parts flying out of the shaper! Hit the deck and the red button!

The forward edge of each tile was being caught by the knives and snagging it enough to where it either broke the forward edge on the vertical grain parts, or took off the entire face on the horizontal grain parts. A high-powered shaper like the SCMI, does not care about the parts flying around—it just keeps cranking. Lots of drama!

We talked a bit about a chip-breaker of some sort. Our motivation was to still make the curve cut in one go. We still had to turn it over and run the back, but that was two passes (x 124 blocks) vs. four passes—another 500 passes! We just did not come up with a way to make the jig work. The photos show the part failure on both vertical grain blocks and the horizontal blocks.

We even joked (seriously) about mounting it on the lathe, but that would be slow. And that grain could still be a problem on a significant enough number of parts to be a no-go. Besides being intimidating to stand near and approach with a gouge...

We thought about a spiral stagger-tooth head, thinking that might minimize the impact. Small chips making lots of cuts in sequence instead of one of two big knives crashing into the part, hammering away until it broke. Time and expense ruled that out.

This jig had some time in it, and our scrap pile was notably depleted by its use of some thick stock. It really was a fine piece of shop work, with density to impart confidence and solidity. It had safety built in, since it could not feed unless everything was in order. The accuracy of the cut was unquestionable. The damn thing just did not work.

Evolution is basic to our problem solving. Complex projects often have to evolve before the path is found. We expect this, but still try to prevent dead ends. This was a dead end since we would have to move to a yet-to-be-devised Plan B. I rarely count on an unknown jig hitting all points on the first go-round. It may even take a third build to get everything needed, incorporated.

We needed to come up with a Plan B for shaping the curved face parts. We did not have one, though my original plan was a linear jig, with the appropriate curve along an edge. Or maybe two. The circular jig won out since we thought it adequate, and very cool.

So, we started in on a straight jig. We would go back to hand feeding, since I felt we needed that feedback and the ability to get out of a cut should it start throwing more than shavings. We turned the feeder around and attached a "foot" to the horizontal beam, and used it as a hold-down, guard, and general security. We used a fixed bearing under the cutter head since we could easily adjust it for flush on our parts. Note in the photo how the curved, fixed bearing was placed into the fences and made for an adjustable "flush" to the straight knives in the head.

The big change was to cut half the face of each block on just the downhill run, not into the grain. This meant each block would have to be set, run, and removed, and run

again, just for one side. Four times, for both sides. Five hundred passes. The jig would hold three parts, so it needed to be loaded 166 times to run the parts. This is what I/we wanted to avoid, but you can't always get what you want…

The evolved jig is similar to the circular Jig of Doom, but "unrolled." The blocks fit onto a lower horizontal tongue and two verticals, a wedge is driven to tighten the verticals, and a one-piece top tongue was fit into place. Handles were added for control, and it was ready to test. The base that the handles mounted to also had the curve times three along the edge facing the cutter.

At times, it is like you pay your dues in one place, and are rewarded in another. The parts came out close to perfect, no scary stuff or minor/major explosions. Even the center match line came out almost unnoticeable. Sanding would be easy. It made no difference what the grain direction was, coming off the high part to the thinner worked just fine. We did not lose a single part.

I continued making blocks and getting them ready for the shaper. Justin spent a day to do the double duty on the shaper, but never complained. The stiles and rails were ready, as simple as they were, but we wanted to wait to size everything until we made up our panels.

With blocks shaped, we could finally see the pattern come up before our eyes as we placed the parts on the bench. It did look woven. Lots of depth to show the shapes, and the shadows said the rest. Very unusual. I have never seen anything quite like it.

We had a fabulous snow come down hard and heavy on a Saturday morning, and the shop was quiet and leaden with four inches of wet snow on the roof. The sun was out from noon on, and cast a remarkable light in the shop. There are times when it is magical in the shop. This was one of those times.

As our shop is small, I generally work several half days during the week, giving Justin full access since he is key personnel, and very productive. If I can help, I'm there, but it is about 50–50 whether I'm helping or bugging him. We will sort through the tasks and assign them to me as appropriate. I have always worked Saturdays and Sundays, as that is my time to advance a project, get caught up, build household or unofficial items, walk in circles and mumble, whatever.

So this Saturday was especially nice since we had a beautiful stack of wonderful parts, the job was back on track, and we had a clear path in front of us. The light was definitely shining on the Acorn this day.

A holding jig was made for sanding with a R/O at 150 grit. It held eight blocks at a time, and in the window light, it was easy to judge accurately what was going on. They were fit onto tongues once again and then clamped to hold in the other direction.

Splines were made and sized to be a loose, not too snug, fit in the blocks. We planned on using epoxy, which likes a thick glue line, and did not need to be

clamped. So I built up a rail to hold the blocks vertically while I buttered the splines with glue and slid them in place—240 splines, that is. They fit well and did not want to creep around as I laid the blocks flat to set up.

We set a straight edge (door stile) and a rail at a good 90° to each other to hold things straight as we glued all the blocks together. We used spacers as place holders where

there was no block edge. Buttered the edges up with epoxy, and laid tile for a while. The fit was just right to allow things sliding together with a tap every now and then.

The hardest part was keeping the alternating grain correct. We watched each other and avoided a calamity. A panel was built on each bench and allowed to cure overnight.

The next day, the panels were rigid, so we left them in place on the benches. No sense tempting fate, eh? The track saw was set up to rip off the protruding blocks on all four sides. A router with a long baseplate was then used to mill a new dado into the edges, all four sides. This would accept a spline—the same size as the ones on the blocks—that would also fit a plow in the rails and stiles.

Splines and grooves were all 3/4 in., as were the tenons and mortises.

These panels were looking a lot like a solid wood panel. The splines were cross grained and would limit some movement. But in my experience, I have not seen Honduras mahogany move. I have 13-in.-wide panels on a west facing door, small overhang over the door in the house, and after 12 years, they finally almost crack the paint. Almost.

Chapter Thirteen: The Business, Innovations, Beadulator, Roper-ator, & Warranties

The Business

This book assumes you are a woodworker with most of the basics in place. It also assumes you are in business, and, as such, need to make a profit, etc. I have run my own business since Acorn was founded in 1990. However, I do not feel I have any particular wisdom to share, despite the fact we have endured numerous recessions, one huge depression, and are now in the midst of a long-term nightmare with the COVID-19 situation. I have limited tricks to beat the crazy things that happen all around us. Mostly, we just knuckle down and carry on, keeping our eye on the project at hand, trying to give it the respect and the time required to carry it through to the best outcome.

So I will skip the basic business advice and offer some observations that, while still basic, are an important part of Acorn's business. First, I describe the paperwork as moving along the path: estimate, proposal, drawings, quote, deposit, and invoice. The proposal is your response to some request for work. The proposal should have everything you need to have in there, and should not have anything you do not want. This is a good time to develop a paragraph that has a line for "Includes" and another for "Excludes." Spell out specifically what will and what won't be done. Detail is not required at this point, but stating the lay of the land is helpful and prevents any early misunderstandings.

Customers come from three main sources: Homeowners often have heard our name and contact us to do work on their project. Builders (building contractors) we know often call us in for various things. Design professionals will specify Acorn in the drawings, or bring us drawings to price and build. At times we are asked to come up with a design similar to what we have done before, or we will be asked to design from the blank page up. Often, there may be expectation or hope from the customer that they can get free design help. This is up to you. I certainly have given away a lot of design help, but almost always with an eye to helping land the project in my lap. If you need to get paid for this bit of design work, be sure you know who to make the invoice out to, and where it goes when it needs to be paid.

The drawings are important. They are based upon field dimensions you have verified and used to make your design—your solution. Clear drawings are your most important communication tool, so they are vital to the clear, concise exchange of information. Be sure to put on your name or logo or some identifier to deter your drawing from being shopped about. Develop a digital letterhead that can go anywhere to ensure that distributed documents are attributed to the correct parties.

Acorn Woodworks
David R. Sochar

Westfield, IN 46074
317-867-4377
davesochar@gmail.com

Customer Pat Redact   **Job** Euclid Ave

Specifications:

Species: Honduras Mahogany (sweitania macrophylla), pattern grade.

Items: Single Door at 1-3/4 x 32 x 86-1/2   4-5/8" wide exterior jamb

Secondary Species: None.

Hardware: 3 4-1/2" hinges, architectural grade, square corner, heavy duty. Without tips. Ball bearing

Latch: By others Mortise latch, add $247.00 Tubular latch, add $180.00

Handing: Right or Left

Sill Type: Oak sill, canted

Weatherstrip: Verticals and head horizontal – Force 5 Compressible foam, with corner blocks. Removable for finishing.

Glass: Code compliant, no logo, clear glass. Single glaze and insulated units are all set with either a light or dark RTV silicone sealant on both sides of the glass. Low E – Add $281.00 plus tax

Exterior Trim: Brick mold is a WM180 pattern, 2" wide. Typically shipped loose and long for field install.

Includes: Shop drawings, Delivery, machine and hand sanding, dummy hinges screws.

Excludes: Latch and latch prep (see above), install, or finish.

As above, Clear glass: $4,546.00 plus tax

Terms are 50% deposit to start work, balance due after delivery.

Fine Architectural Millwork
Bench Made Furniture

The quote is the formal description of the project, with all the details you need to repeat back to the customer.

It will have the specifications and the pricing, with options and exceptions. Terms should be spelled out. We have always invoiced for 50% when the proposal is accepted, then we make the quote and then the deposit invoice. Being lazy of a sort, I merely lift the proposal language and paste it into the quote and deposit, making any needed changes along the way. I use QuickBooks like most everyone else, so text can be lifted from a word processor to the quote page with a couple of clicks.

Acorn Woodworks
16116 Ditch Road
Westfield, IN 46074

317-867-4377
acornw@frontier.com

Quote

Quote No.
1719

Name/Address	Date	Terms
Booker T Sample The MG's Green Onion, OK 32245	07/19/20	
	Project	P.O. No.

Item	Description	Total
Entry System	Main Entry - To be paired 2-1/4" x 36" x 120" doors per drawings attached, based upon the architecturals as provided. The top of the doors and frame is to be a segment arch, with a radius derived proportionately. The doors are to be solid Pattern Grade Honduras Mahogany (swieteneia macrophylla) for beauty, durability and long life. The doors are 3 panels each, with each panel being retained by a large Bolection molding. The bolection mold is 'landed' on a continuous spline to reinforce the miters and make the molding assembly solid and stable. The matching frame will be with a sawn true radius curve for the head, with a solid liner. A 2" WM180 Brick mold will be shipped loose, for install on the job. All silicone bulb perimeter weatherstrips will be provided loose, and the Oak sill will be sized for finish floor thickness and a 1" allowance will be made for an interior rug. We will fit and hang the doors on 4- 6x6 ball bearing, ball tip hinges per door, with extension bolts to retain the passive door. Mortise latch prep is included, if the latch is provided to Acorn during fabrication. All work will be fully machine and hand sanded to 150 grit R/O, and ready for finish. Includes: Site verification, shop drawings, doors, t-astragal, frame and sill, hinges and extension bolts, latch prep, weatherstrip, exterior trim, sample for finish. Includes local delivery. Excludes: Mortise latch, interior trim, finish, installation. Sales Tax	18,000.00T 0.00
	Total	$18,000.00

Acorn Woodworks

Invoice

Invoice No.
1663

Bill To
Booker T Sample
The MG's
Green Onion, OK 32245

Date	P.O. Number	Terms	Project
07/19/20		Due on receipt	

Item	Description	Amount
Entry System	Main Entry - To be paired 2-1/4" x 36" x 120" doors per drawings attached, based upon the architecturals as provided. The top of the doors and frame is to be a segment arch, with a radius derived proportionately. The doors are to be solid Pattern Grade Honduras Mahogany (swieteneia macrophylla) for beauty, durability and long life. The doors are 3 panels each, with each panel being retained by a large Bolection molding. The bolection mold is 'landed' on a continuous spline to reinforce the miters and make the molding assembly solid and stable. The matching frame will be with a sawn true radius curve for the head, with a solid liner. A 2" WM180 Brick mold will be shipped loose, for install on the job. All silicone bulb perimeter weatherstrips will be provided loose, and the Oak sill will be sized for finish floor thickness and a 1" allowance will be made for an interior rug. We will fit and hang the doors on 4- 6x6 ball bearing, ball tip hinges per door, with extension bolts to retain the passive door. Mortise latch prep is included, if the latch is provided to Acorn during fabrication. All work will be fully machine and hand sanded to 150 grit R/O, and ready for finish. Includes: Site verification, shop drawings, doors, t-astragal, frame and sill, hinges and extension bolts, latch prep, weatherstrip, exterior trim, sample for finish. Includes local delivery. Excludes: Mortise latch, interior trim, finish, installation.	9,000.00T

Deposits	$0.00
Subtotal	$9,000.00
Sales Tax (0.0%)	$0.00
Balance Due	$9,000.00
Total	$9,000.00

The invoice is sent when the project on that invoice is completed. By this time, I know I can trust my customer. For many, I will put terms of 15 days just to expedite the payment. Longer than that, and I may feel it on the bank account end as it gets too light. It is rare to ever have to chase a payment from a deadbeat customer. The bank crisis of 2008 caused the bankruptcy of a builder, and I will never see that money. But in 30 years, I have lost less than two weeks' income. I do not choose to work at the bottom of the market, and the lack of of problems getting paid is indicative of the fact that our customers can afford our work. Splitting hairs, figuring cheap and cheaper, is a losing game, as you just forever chase an absolute that is unapproachable. I will gladly let others chase the 422 doors, or the 32 staircases.

This path is easily tracked by QuickBooks. If you do not have it now, I suggest you set it up with the advice of a QuickBooks expert in your area. Easily found through the QB website, an afternoon will help you get it set up right, and you will get all the benefits QB offers. If you use an accountant or CPA, then it will be mandatory that you supply info to them at least on an annual basis for tax preparation. You will come to depend upon the QB product for its reporting, bill-paying and tracking of all things.

Once the proposal is accepted and the deposit made, the paperwork to build the project is completed. We use two simple forms, one for the door(s) and one for the frame. These are forms we made, on the fly, with the originals about 30 years old. Needless to say, they have evolved somewhat, and now are pretty thorough at collecting all the needed info into one place. But remember, this is custom work, so it will be impossible to make a document that that will catch everything. Perhaps a blank sheet of paper.

Door Build Information

Drawing for this item attached Yes____ No____

	Door QTY	Sidelights QTY	Transom QTY
Size	Th___W___H___	Th___W___H___	Th___W___H___
Species	H. M. Other_____	H. M. Other_____	H. M. Other_____
Description	_____	_____	_____
Sticking	9/16" 11/16" ___	9/16" 11/16" ___	9/16" 11/16" ___

Panel Th. _____ Panel Raise Hip Ogee Cove

Other_____

Horizontal _____

Vertical _____

Glass Thick_____ Acorn Set Yes No Seal Gry Blk Black Stops Yes No

Glass Dimensions _____

Glass Supplier _____ Glass Template Yes No

Fit and Hang Yes, on Frame Sheet No

Install Sash w/3/16" quirk Yes No

T-Astragal Yes No

Special Instructions _____

Customer _____ Job _____

Date Due: _____ Budget Hours: _____

Each project gets one each of the frame and door builds, and any project that has three to four distinct projects will get three to four of the frame and door builds. Drawings will also be included in this package that goes out to the shop. The package of documents should have everything the woodworker needs to build the project.

Exterior Frame Build Information

Drawing for this Frame Attached Yes____ No____

Tag this Frame with "Set This Frame..." Yes____ No____

Species: H. Mahogany
Other:_____

Sill: White Oak, canted, 1-3/4"
White Oak, flat, 3/4"
Aluminum, dark bronze
Other _____

Jamb Width: 4-9/16" 6-9/16"
Other_____

Rabbets: 1-13/16" 2-1/8" Q-lon
2-5/16" 2-5/8" Q-lon
Other_____

Brick Mold: 2" Loose or Attached
2-3/4" Ogee Loose or Attached
4" Ogee Loose or Attached
Other _____ None

Opening Dimensions: Door(s)

Thick_____ Width_____ Height_____

Sidelights: Width_____ Height_____

Transom: Width_____ Height_____

Inside Rabbet Head Radius _____

Exterior Mulls: Yes No
Other_____

Interior Mulls: Yes, loose No
Other_____

Outside Frame Width: _____

Top of Sill to Top of Head Dimension: _____

Horizontal Frame _____

Layout: Sash _____

Vertical Frame _____

Layout: Sash _____

Rough Fit or **Fit and Hang** Left Hand Right Hand Right Hand Active Left Hand Active

Hinges 3, 4, 5 @ 3-1/2" 4" 4-1/2" 5" **Bottom Clearance** 3/8" w/ Epoxy, rabbeted Yes No

Latch Preparation Yes No **Brass Interlock** Yes, w/ nails and screws No

Pairs Prep Flush bolts Ext Bolts **Astragal:** Yes, Q-lon Yes No Astragal

Customer: _____ Job: _____

Date Due: _____ Budget hours: Build_____ Fit/Hang_____

These forms were made years ago and generally do the job. I often think of reworking these, but they are familiar. Actually, familiarity is important in forms—one learns where the hinge count is, and there it is every time. If it is not there, then it has not arrived yet. The forms are a useful tool for keeping it all organized. Learn to use the forms, no matter if the project is large or small, enter it all in QB and you almost can't go wrong. Well, at least the hard part will still be the woodwork. The paperwork is easy when you stay on top of it.

Innovations

In an effort to set our business above the others, we have sought out the best materials and methods, occasionally developing new or improved methods. This should work to our advantage in convincing potential customers that this company has everything they need for their project. Successful innovators can place themselves in the market where they like and enjoy watching others try to catch up. To that end, we have developed a few things, large and small, that should help set us apart from our competition.

First, epoxy end coats. We learned to coat the bottom of every door we fit and hang with epoxy. This will prevent water absorption for 40 years or more. The material soaks up into the end of the stiles about 1/4 in., and blocks water absorption. This preserves the stile bottom rail joints that are often just opening up on new doors after a few months in service and several rainfalls. This simple but important bit of work is something painters rarely do even though every door-maker's warranty will prescribe full paint on all six sides of the door, including hardware mortises.

Second, dummy screws along with our hinges. The screws packaged with the hinges are brass, and soft enough to twist off with power drills. Especially when the screw is not pre-drilled. The bagged and unlabeled screws help the astute carpenter as he hangs the door, and the job will benefit since he is able to install the proper screws as he is finishing up. However, in about half the jobs we inspect, the dummy screws are never used. The brass screws are missing from the hinges, or snapped off and glued back in, or rounded out, or colored with a magic marker, or worse. Observing, we can find no reason why these screws are not used. Upon asking, we get a blank look. We are developing an instruction sheet to inform the user what the screws are for, and how best to use them. Printing on brightly colored paper might help get their attention. We just want to help. But so many carpenters are undertrained and overworked, never having had the proper training or even exposure to good craft. Throw in the competition that is every carpenter's burden, and defensiveness is the response.

Third, brick mold template. The most visible innovation we have developed is the brick mold template. This is merely a panel of OSB that is the exact size of the exterior finished entry with trim. It has our logo spray-painted by stencil at two feet square, so anyone onsite or driving by knows we are supplying the doors. We used to charge $150 to $250 for each one, but now we just include it in all new work. It is in the estimate, but not a line item on any invoicing.

It has been an uphill battle trying to get builders to use the brick mold template. In this market, the folks who sold the Vietnamese and Indonesian entries by the container-load sent the entry out during rough framing, then they would set the doors, and then return them to the warehouse for storage until finish carpentry. The frame and exterior trim would remain. The brick, stone, wood, or other exterior cladding could be run right up tight to the trim, and be complete.

Meanwhile, the frame took the brunt of the weather and dangers of the job site. They were often dented, grayed out, splintered, and worse. A single sheet of Visqueen would be flapping in the breeze, protecting nothing. The jamb would require quite a lot of attention to get cleaned up enough to be presentable.

When we started shipping entries, we had to explain that we were doing it differently. With the brick mold template, we could ship the entries anytime, and they could be set by about any carpenter. The template will close the building up and keep the varmints out. A temporary or cull door can be set if needed. Insulation can be added to the inside, and the opening will be cool/warm. This also gives us the luxury of time. We have time for the customer to make up their mind. They can wait until interior trim is imminent before deciding. This also gives us more time to place the job in our schedule. We have jobs that ship in as little as four weeks from first contact, to two years form the first call.

To recap: more time for owners and builders, no fine woods out in the weather, no delay in finishing exterior cladding, a closed-up building for security, no additional costs, etc.

The template includes the brick mold and the bottom of the sill. We ship the brick mold loose on our entries, so the carpenter has the ability to fiddle a bit if needed. Carpenters can set the frame, then set the trim and doors, and get everything trued up and operable. Then, they can pick up the template and put it up and over the entry to protect it. Again. Our logo will remain intact, and the entry is protected. All in all, one of our better ideas.

Beadulator

I drew the short straw on a large job of interior doors that all had a 5/8-in.-diameter bead running around each panel. This was no weenie bead—a full half-bead, almost 7/16 in. proud, and 550 L/F of it! But the trouble with a mold like this is the repeat. Making the molding is one thing, but being able to manipulate the spacing in order to have a bead come out exactly in each corner would be difficult. The machine would have to somehow make the spacing a bit variable.

The drill press received a heavily ground upon flat bit that was to form the bead. The grind was tedious to create a half-round bead that would be sharp and not make any tear-out. The alder molding was 7/8 in. wide by about 5/8 in. thick in lengths up to 7 ft. or so.

The spacing adjustment could be minor, less than 1/16 in. since it could be deployed a little or lot at a time. I developed a marking line of 13/16 in. on the drill press, and a similar series of lines on the last foot of the molding blanks. We had about ten 9-ft. doors, each with about 48 L/F of bead required.

The drill press required a careful hand: feed too fast and the beads could twist off. Feed too slow, and they would burn. I did get pretty good at adjusting the spacing as needed. And once the three days of set-up and run were over, the darn things looked great when they were laid into the moldings. As soon as the run was completed, the temptation was there to tear down the set-up on the drill press and return it to normal service. But I left it up—just in case—and indeed had to go back and make more molding, as one or two pieces were still required to finish the job.

Once sanded and finished, these were some of the most attractive doors I have made—simple and stately. I am not a big fan of alder, but the finish was very well done, and they came out exceptionally nice. Most of them went to pairs at 9 ft. tall, with distressed glass and a low horizontal panel, with cremone bolts to hold the pair closed.

A word about the beads and corners. It is important to match carved or other decorative elements at corners. Rubber, plastic, or pre-carved elements give away their simple heritage when they butt to each other or come to corners. Part of the carver's skill is anticipating corners and how the elements will match at the corners and proceed along to the next corner. Even though most observers will not know to look, I feel it is imperative to recognize, respect, and repeat this nice touch. The beads to the right would not look so nice with partial beads at the corners.

Roper-ator

The Roper-ator is a temporary device built to make curved rope molding. It consists of a base that will hold the curved molding blanks, a circular (or is it tubular?) fence of sorts to help guide the router, an indexing stop, and some clamps to hold the part. The parts were 1⅝ in. in diameter and mated up to straights that we were making with with a legacy mill. The inside radius was about 14 in., and each part was just over 90° in "length." The rope profile did not have to go all around the part. A small P-C router was used and loaded up with basically a V-pointed cutting tool. A smaller, tubular fence was fashioned to support the router close to the workpiece, and the larger helical part gave us all that we needed. A careful, slow and steady pace kept it all together. In the center was a DESTACO clamp that held the molding blank in place. The index stick located the blank for each cut by catching the previous cut. The molding cut was only on about 270° of the blank—it did not go all the way around.

The application was three tall narrow doors with half round heads. Straight "rope" ran on both sides of each door as part of the casing, but stopped at the springline. We were to make six 90°-plus arches, with the twist opposing and meeting at the head, at the top center. Any butt of the casing had to continue the rope twist. The catch was, they had to mate up to the straights at one end, and then meet correctly at the top so they made a V at the center.

We started with about 14 blanks in poplar. Thinking we'd need to ship six, the eight extra would come in handy. We used every bit of them, and shipped the best six. The routing became simpler as we went, and we learned how to monkey with the indexer to end up where we needed to.

Warranty

The Acorn Warranty is five years, four years longer than most. This should help draw business, since a longer warranty is greater protection, especially for a wood door. We have had this warranty for about 12 years, and we had several claims when the panel glue started failing all over town. We stood behind every door, and made good on the warranty. That is the problem with a warranty: If you put it out there, you better be prepared if there are claims. Fortunately, we have had no claims since the panel failures. I could not even think about a warranty unless I believed 100% in our product.

Research other makers' warranties and you will see that they all have conditions in common. No one strays too far from the core message of the warranty: "If it is a manufacturer's defect, and you can prove it, and it matters." One employer told me I needed to write a door warranty that no one could ever make a claim on. I did it since I had been shown the ultimate loophole on door warranties some years prior.

The warranties make it clear that the finish—paint or stain—must coat all surfaces. "All" means inside hardware prep, door bottom, door top, etc. As literal as possible. If these areas are not properly addressed, it could void the warranty. An old dodgy guy once told me that the way one skated out of the warranty was to take one hinge loose from the door to see if there were three, four, or five coats of finish inside the hinge mortise. There almost never were, and this is where the warranty is voided. If by chance that inside hinge pocket was finished properly, then look under the latch hardware, especially inside the bores. Chances are, lack of finish in this area will allow voiding the warranty.

I have heard that this is what door company reps do. They do not examine the door for warped, panel splits, or veneer lifts. They go straight to the hinge, take it loose, photograph it, and move on. Even though it is obvious that the lack of a couple coats of finish in a hinge pocket would not cause a panel to split.

Not that it is the goal of any proper woodworker to make things to avoid warranting them. Our goal is to make products that exceed the warranty with ease. To never need the warranty or to build so close to the warranty that we have attorneys police it for us. As we know, small businessmen have plenty of ethics tests placed before them. Learning to work well and honestly is part of being in business.

Index

ADA thresholds, 131
air. See ventilation
air, compressed, 19
Angel-rig, 148
astragals, 154

battens, 136
Beadulator, 181–83
bearings, 71
bench (shop), 13–15
board and batten, 162
board footage, 4, 9
bolts, 150, 152
bolts, Cremone, 153, 182–83
brick-laid heads, 146, 148
brick mold template, 180–81
brushes, 66
bumpers, 57, 66
bust-out, 11, 36–37

carbide chips, 40, 56
carpenters, 43, 152–53, 154, 157, 181
casing, 123
caulk, 90
chamfering, 114
chisels, 25, 56, 64
chords, 146
circular saw, 57
clamps, 58, 67, 121, 136, 137

clavos, 88, 108, 152
clearance, 63, 111, 117, 128
climb-cut, 24
code (building), 91–92, 93
color matching, 30
compound curves, 123–24
compressible foam, 115, 128, 149–50
compression, 53, 159
cope and stick, ix, 50, 52, 71
cope cutters, 70
copes, gluing, 72–73
coping, 50, 52
corners, 183
costs, 3–4, 21, 23, 27
costs, indirect, 5–6
cracks, 84
cross bands, 30, 112, 114
cross-section, 6, 7–8, 78, 98
crowning, 43, 115, 150–51
crush zone, 88, 111
curve cradle, 122
curved heads, 139–43
curves, sawing, 137–38
customers, 11, 176, 178
cut lists, 7, 8, 11, 36–37, 61
cutters, 40, 52, 54, 55, 56, 70–71

dado saw, 103
digital calipers, 42, 44

digital readout (DRO), 42
dimensions, format for, 37
divided lights, 76, 95, 97
doors, types of
    bolection, 77–80
    combination, 115–16, 117
    coped, 68–75
    Dutch, 118–19
    exterior, 34–35, 73, 77
    flush, 111–15
    frame and panel, 81
    large, 124–25
    overhead, 163–64
    passive, 154
    pivot, 115
    plank, 81–82, 107–11
    sashed, 75–76
    screen, 117
    secondary, 115–18
    simple, 61–62, 64–68
    storm, 116, 117
dowels, ix, 73
draw boring, 52–53
drawings, 6–9, 11, 133, 176
    CAD, 7
    cross-section, 6, 7–8, 78, 98
    elevations, 7, 133
    full-size, 136, 139, 140, 141
    hand, 7, 8
dry fitting, 62, 64–65, 90

dummy screws, 154, 180
durability in service, 34
dust collection, 16–18

egg crate core, 113, 114
endless board, 25
epoxy, 30, 59, 68, 74, 86, 87, 114
epoxy end coat, 74, 82–83, 151, 180
equipment, 20–28
estimates, 3–6
expansion, 62–64, 67, 77, 88
expansion gap, 109

facing, 37–41
feeders, 24–25
fences, 24, 138
forms, 3, 179
frames, 127, 145–48
free of heart center (FOHC), 161

gaining, 68, 69
gap-filling, 59
gates, 49, 53, 159–62
glass, 89–94, 144–45
   types of
      art, 93–94
      insulated, 92, 144
      laminated, 92–93, 116
      safety, 91–92
      tempered, 92, 93, 116
glazing, 95
glue, 58–59, 67–68. *See also* epoxy

failure of, 29, 84
removing of, 74
surface area of, 49, 50, 52, 53, 65, 73, 79, 85
gluing, 57, 66–67
gluing for width, 29, 30, 43, 52, 83
grain, 140, 167, 172
grain matching, 30, 86, 124
green tally, 10
guarantee, 8. *See also* warranty
guarantee of payment, 9

hand grind, 72
hanging, 130, 150, 153, 154–55
hardwoods, 10
haunching, 65–66
hearing protection, 38
heat, 29, 116
height (in shop), 14, 19, 20
hingeing, 154–55
hinge-routing template, 155
hinges, 117, 125, 156–57, 162
hinges, gate, 159, 160–61
hold-downs, 90
hunting miters, 139
hygroscopic movement, 63

index, 100, 184
insulation, 110, 130, 181
insulation value, 93
invoices, 177, 178
isotropism, 62

jambs, 127, 129–30, 145
J-hook, 151, 152
jig, tenon, 26, 27
jigs, making, 99, 128, 167, 170
joiners (jointers), 22, 23, 37, 38–39, 41
joinery, 47–53
joints, 47, 48, 49, 83–85, 136

Katz, Gary, 139, 150, 156
kerfs, 115, 129, 147, 150, 151
kiln-drying, 10, 109
knife cuts per inch (KCPI), 72
knives, 40, 51, 56, 71–72, 83

labor, 4, 31. *See also* shop rate
ladder core, 109–11
lamination, stack, 52
lands, 76, 79
latches, 118, 119, 157–58, 161
layout, 43–45
leakage. *See* water
leveling, 68, 74
lifting, 19–20
lighting, 15–16
loading dock, 19
logos, 92, 116, 181
louvers, 97–106, 163
   types of
      curved head, 102–6
      false, 101–2

lumber, 4, 9–11, 33–37, 161–62
  density of, 34–35
  grades of, 52
  species of, 33–34, 36
    hardwoods, 10
    Honduras mahogany, 35–36
    pine, 33
    poplar, 33–34, 36
Lyon mold, 77

mallets, 57
marking conventions, 37, 44–45
markup, 4
MDF, 30
metalwork, 162
mills, 10
miters, 51, 79
moisture content, 10, 62–63, 109
molding, curved, 135–39
moldings, bolection, 57, 75–76, 77–80
moldings, stand up, 138
mortise and tenon joints, 48–49, 52–53
mortisers, 25–26
Mouzon, Stephen, 97
mulls, 129, 130, 131
muntins, 50, 95–97
muntins, curved, 143–45

nailers, 91
net tally, 10

oriels, 162
overhead (financial), 5
ovolos, 50, 68

paint, 90, 91
  clearance for, 117
  door bottoms and, 74–75
panel assembly, 106
panel plows, 50–51
panels, 79–80, 81–89
  types of
    fielded, 81
    raised, 81, 88
    wide, 84
    wood, 82
    woven wood, 89, 165–74
pickets, 159–60
planer, hand, 57
planers, 23–24, 42
planing, 42–43
plows, 65, 66, 103
pneumatic closers, 118
posts, 161
process, 15, 20, 28–31
profiles, 54, 55–56, 96, 143
profit, 5
proportion, 139–40
proposals, 5, 11, 175–76

QuickBooks, 177, 178
quotes, 177

rabbets, 56, 89, 117, 128, 129, 132, 146, 148
racking, 48
radius, 103, 121, 136, 146
radius plan door, 120–23
rails, 45, 61, 67, 69, 103, 139–43
raises, 82
ripping, 37, 42
rippings, 57, 61, 67, 97, 111
Roper-ator, 183–84
roughing out, 37
router guides, 100
routers, 26, 101, 121, 142, 151, 152–53, 155

safety, 21, 39, 89–90, 142–43
sanders, 19, 22
sand-throughs, 112
sash bars. See muntins
sashes, 75–76, 91, 117, 132–33
saw, circular, 57
score.org, 6
scrapers, 68, 74
sealant, 80, 90, 131, 132, 133
  glass and, 50, 79, 91, 93, 116
seasonal changes, 30, 63, 67, 102
shapers, 24–25, 27, 56, 138
shims, 14, 55, 70
shipping, 93, 151, 181
shop furniture, 13–15, 19
shop rate, 4–5, 6

shop space, 13–20
shrinkage, 63, 64, 88
Shrinkulator, 63
shutters, 162–63
sidelights, 129, 130, 131, 132, 133, 151
silicone bulb, 115, 129, 150–51
silicone RTV, 90, 91, 133
sills, 130–33, 151, 152
site assembly, 130
slats, 98, 100, 101, 105, 106, 163
spacers, 55, 70, 97, 109, 129, 144
specifications, 9, 11–12, 176, 177
spline, 79, 84, 87
spokeshave, 137, 142
springline, 141
squares (combination), 57
squeeze-out, 67, 72–73, 91, 131
steam bending, 145–46
sticking, 50, 54, 68, 69, 70, 89, 121
stiles, 36, 43, 44–45, 100

stops, 91, 127, 136
surface four sides (S4S), 42–43
sustainability, 35

templates, 136–37, 142, 155, 180–81
tenon discs, 27
tenoners, 26–27
tenons, 26, 27, 29, 48, 53, 65
tension, 159, 161
terms (invoice), 177
T-horses, 20
threshold cap, 151
tongue and groove boards, 86, 88, 111, 164
tooling, 54–56, 71
tools, handheld, 21, 57–58
trammel sticks, 136
transoms, 130, 133
tree nails, 107
tree rings, 34–35
two sticks, two lines, 14, 133–34

vacuum bags, 111, 114, 122
vendors, 9, 10–11, 92, 93
veneer, 30–31, 85–86, 111–13
ventilation, 16–18
V-joint, 86, 111

warranty, 36, 84, 91, 145, 180, 184–85
waste factor, 9–10
water, points of entry for, 73, 74–75, 77, 79
water repellency, 51, 55, 82–83, 133, 151–52
weather, 33–34, 59, 75, 79, 116, 151
weatherstrip, 115, 117–18, 128–29, 149–51. *See also* sills
wedges, 53
wind, 101, 118, 160
wood, properties of, 62–64

zig-zag stitcher, 113

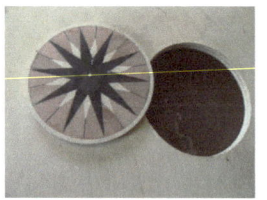

www.ingramcontent.com/pod-product-compliance
Lightning Source LLC
Chambersburg PA
CBHW041820080526

44589CB00004B/61